PLANTS & GARDENS

BROOKLYN BOTANIC GARDEN RECORD

DYES FROM NATURE

1990

Brooklyn Botanic Garden

Staff for this Issue:

Rita Buchanan, GUEST EDITOR

Barbara B. Pesch, DIRECTOR OF PUBLICATIONS

Janet Marinelli, ASSOCIATE EDITOR

and the Editorial Committee of the Brooklyn Botanic Garden

Bekka Lindstrom, ART DIRECTOR

Judith D. Zuk, PRESIDENT, BROOKLYN BOTANIC GARDEN

Elizabeth Scholtz, DIRECTOR EMERITUS, BROOKLYN BOTANIC GARDEN

Stephen K-M. Tim, VICE PRESIDENT, SCIENCE & PUBLICATIONS

Cover Photograph by Rita Buchanan

Plants & Gardens, Brooklyn Botanic Garden Record (ISSN 0362-5850) is published quarterly at 1000 Washington Ave., Brooklyn, N.Y. 11225, by the **Brooklyn Botanic Garden, Inc.** Second-class-postage paid at Brooklyn, N.Y., and at additional mailing offices. Subscription included in Botanic Garden membership dues ($25.00 per year).

ISBN: 0-945352-58-1

PLANTS & GARDENS

BROOKLYN BOTANIC GARDEN RECORD

DYES FROM NATURE

VOL. 46, NO.2, SUMMER 1990

HANDBOOK # 124

PHOTO: METROPOLITAN MUSEUM OF ART

The Unicorn Tapestries were woven about 1500. The colors were made from dyes from plants.

FOREWORD

RITA BUCHANAN

Using dyes from plants dates back thousands of years in all societies around the globe. The first step was identifying potential dye plants; then came learning how to extract the pigments and apply them as dyes. Primitive dyeing techniques were often very elaborate, as the goods to be dyed were alternately treated and dried over a period of weeks or months. Yarns or fabrics were soaked, dipped and coated with a variety of different substances. All this was necessary to achieve rich and durable colors.

Dyeing was well established as a skilled craft among the ancient civilizations that surrounded the Mediterranean — the Phoenicians, Egyptians, Greeks and Romans. In those cultures, although surely some people dyed their own yarns and clothing at home, most dyeing was done in special workshops equipped with big vats and sinks and drying racks. To the East, dyers from the Caucasus to Persia developed techniques for imparting beautiful vivid colors on wool and silk yarns for weaving their elaborately figured knotted rugs. Seen in museums today, many of these Central Asian tex-

RITA BUCHANAN, *the Managing Editor of* ***Fine Gardening*** *magazine, is the Guest Editor of this Handbook.*

tiles are still distinctly colorful; they must have been even more remarkable when they were made hundreds of years ago.

By the Middle Ages in Europe, dyers were well organized into professional guilds, with prolonged training periods for apprentices and high standards for master dyers. The guild system was based on division of labor; growing, processing and using dye plants were separated as different trades. Guilds in different regions often specialized in the use of one or another dyestuff or the dyeing of different fibers. There was considerable rivalry and jealousy between guilds, and dyers frequently attempted to infiltrate each others' workshops and steal trade secrets.

In the cities and more populated regions of the colonial United States, dyeing was again the business of trained craftsmen, many of whom had learned their skills in European workshops and depended on imported dyestuffs. Dyers had an increasing knowledge of chemistry, and began to experiment with different mordants and dyebath additives. In both Europe and the United States, books of dye recipes were written and published for the first time during the 1700s and early 1800s. At the time spinning and weaving were becoming increasingly mechanized, so more and more cloth was being produced. Commercial dye workshops had to handle large quantities and produce consistent results.

A synthetic dye pigment was one of the first products of modern chemistry. In 1856, eighteen-year-old chemistry student William Perkin accidentally discovered a coal-tar derivative that made a vivid purple solution. Curious, he tested it as a dye on silk and had his results confirmed by a major dye house. Perkin was in business within a year, manufacturing the first synthetic aniline dye, which he called mauve. His business was so successful that he retired at 35 to return to doing research.

Other chemists, especially in Germany, quickly began to study the structure of natural pigments and tried to duplicate them. A process for synthesizing alizarin, the red pigment in madder roots, was patented by 1869. The blue of indigo was described in 1880, and synthetic indigo was marketed after 1897. Although the first synthetic dyes tended to run or bleed into wash water and to fade in the sun, they were bright, inexpensive and very fashionable. Within a few decades after the discovery of synthetic dyes, the market for natural dyestuffs collapsed.

Dyeing with plants enjoyed a rebirth of popularity in the United States in the 1960s and 1970s. Two previous Brooklyn Botanic Garden handbooks — *Dye Plants and Dyeing* in 1964, and *Natural Plant Dyeing* in 1973 — were welcomed by dyers at a time when there were few references to study and few authorities to consult. The 60s generation of enthusiastic and inventive dyers developed many new approaches, rediscovered some old ones and tried all the plants they could gather. Most dyeing was done on wool yarn, and soft muted earth-tones were popular colors in those days.

Continuing interest in natural dyes appears in many ways. Textile hobbyists in developed countries prefer the appealing colors of naturally dyed yarns for their knitting and weaving projects. Craftspeople in underdeveloped countries use natural dyes for economic reasons: The export value of handmade textiles increases if they are dyed with natural rather than synthetic dyes. Dedicated researchers and scholars are discovering and documenting the techniques of dyers long ago and far away. With all this support, it appears that the ancient traditions of natural dyeing with plants will be carried into the 21st century. ❖

COCHINEAL
THE BUG IN THE RUG

GARY N. ROSS

My passion for cochineal began in 1978, when I visited the Zapotec village of Teotitlan del Valle. Most of the men in this semi-arid highland community of more than 5,000 are weavers, and the village is famous throughout Mexico and among connoisseurs worldwide for its hand-loomed tapestries and rugs colored by fine natural dyes. I entered the showroom of Isaac Vasquez, then 43 years old, whose work I had seen in an art gallery in San Antonio. I later learned that he was the most prestigious master weaver and dyer in the contemporary Zapotec culture. My host expounded in slow, deliberate Spanish — his second fluent language — "These colors can only be achieved with natural dyes, and only I and my cousin Alberto Vasquez, whom I taught here in Teotitlan, are willing to take the time to prepare the dye bath as our ancestors did so long ago."

My eyes fell on the profusion of reds — crimson, scarlet, maroon, mauve, coral, pink and lavender. Isaac noticed my wonderment. "Yes," he continued, "the reds are especially beautiful and my favorites, too." Then he added: "Did you know that we Zapotecs cultivated the dye-producing bug, the cochineal insect, long before the arrival of the Spaniards?" I had to admit my ignorance. The weaver walked to a small table, picked up a jar of what looked like small, gray seeds, and handed it to me. The seeds proved to be the dried bodies of thousands of tiny insects. Isaac explained that the dye is relatively simple to prepare. The brittle bugs "are ground into a powder and boiled in a water solution with lime juice and certain dried leaves."

I was hooked. Each day of the following week I returned to Teotitlan del Valle and sat for hours observing the various weavers at their looms. Isaac became impressed with my persistence. One afternoon he looked me directly in the eyes and asked: "Would you be interested in documenting the Zapotec method of dyeing?" He explained that he had gleaned the cochineal recipe from the village elders when he was still a boy, and that he had not divulged it to any outsider before. I was honored but told him that for the time being I had to return to Louisiana to resume my teaching and research duties.

Cochineal Natural History

During the next three years, I researched cochineal dye in my spare time. I learned that entomologists classify the dye insect

GARY N. ROSS *is a professor of biology at Southern University in Baton Rouge, Louisiana. He has conducted research on insects, handwoven textiles and natural dyestuffs, and Indian cultures in Mexico and Central America for over 25 years.*

in the order Homoptera, along with the more typical mealybug and scale insects that plague many house and garden plants. The species name of the bug (now *Dactylopius coccus*) has varied considerably throughout its illustrious past, primarily because there are two distinct forms. A small, wild cochineal lives on prickly pear and similar cacti, which are found from Florida to Arizona and New Mexico, and south into the drier parts of Mexico and Central and South America. The slightly larger form, the cultivated or domesticated type, depends on human care for seasonal protection from the cold and thrives specifically on cacti of the genus *Nopalea,* called *nopal* in Mexico. Both forms of the insect produce the brilliant red dye, but the domestic form, generally preferred because it is larger, is the only one that is commercially available.

The unusual natural histories of the two forms of the insect are virtually identical. Young females settle near their mother in groups of four to ten and secrete from their abdomens a white, weblike material, which accumulates and becomes a cottony puff. This wax-based secretion, more profuse in the wild form, is an effective barrier against desiccation in the semiarid climate and a camouflage against potential predators such as birds, small rodents and other insects. Within her protective cloak, each female is tethered to the cactus pad by a tubelike proboscis. Through this tube she draws nutrients from the host plant for her entire life, up to three years. Within a few weeks of hatching, the female cochineal becomes bloated and sexually mature — a silvery, purplish-black balloon about one-quarter inch long. It is these females that are the source of the dye: A deep maroon pigment pervades their body fluids and tissues.

The male cochineal remains diminutive and mobile. After a few weeks, he spins a cocoonlike structure and transforms himself into a tiny reddish-brown, soft-bodied, delicate-winged flyer. Lacking mouthparts, the male is instinctively programmed for only one thing, sex, and that he pursues with relentless vigor, moving rapidly among the cactus pads and copulating with as many females as possible. The adult male dies after about a week.

The numerous eggs deposited by the female hatch immediately into tiny, reddish-brown, six-legged specks. Those specks that will eventually become mature females usually remain on their home turf, puncturing the plant's cuticle, or wax layer, in order to get down to the business of feeding and reproducing. Inevitably, however, as the season progresses, the cactus-pad home and nursery becomes overpopulated. No matter. The newly born babies move easily on to greener pastures — an adjacent pad — and set up new colonies. When all the home sites on a plant are taken, the tiny creatures cast their fates to the wind, the successful ones making a suitable landing on a fresh and uninhabited cactus.

Dyeing with Cochineal

Equipped with this background, I returned to Oaxaca to document the insect's use as a textile dye. Isaac wasted no time on my initiation: The day after my arrival we undertook an arduous drive to the lush Pacific lowlands to collect the leaves of the *tejute* or *hoja lisa* (*Miconia argentea*), a medium-sized tree with an umbrella canopy and large, coarse leaves with whitish, heavily veined undersides. After stuffing four large grain sacks with leaves, we backtracked to Isaac's home. Then, while the *tejute* leaves we had obtained dried in the highland's potent sun for several days, my host, the "keeper of dye secrets," filled me in on his art.

Most activities in a Zapotec household are concentrated in the central, open-air

The small cochineal insect feeds on a cactus called nopal, *Nopalea*.

patio. The Vasquez patio has one small area reserved for dyeing, a shaded spot protected by a roof overhang. There a fifteen-gallon aluminum pot rested on a simple tripod of rocks. Stacked nearby was wood gathered from the mountains above the village. First, the pot had to be filled with well water and the fire kindled to bring the contents to a boil. About two hours of heating were required, during which time Isaac and his wife, Maria, prepared the wool and ingredients for the eventual dye bath.

Isaac took three skeins of his wife's handspun woolen thread, about two pounds and soaked them in a tub of clear, cool water. As soon as the dyepot started steaming, three to four handfuls of the dried, crumbled *tejute* leaves were added. My guess is that the leaves contain oxalic acid, as do the leaves of some other members of the plant's family, Melastomaceae. This acid is recognized in the professional dye literature as a color intensifier and mordant for cochineal. (A mordant is a substance, usually a metallic salt, that combines with the dye to form an insoluble colored compound on the fibers of the yarn.)

As the water continued to heat, Isaac plucked approximately eighty fresh limes from a tree that grew conveniently in the middle of his patio and squeezed out the acid juice by hand. I questioned Isaac about his use of citric acid; the citrus family of plants is considered a native of the Old World, having first entered Mexico during the early colonial period. "The Zapotec's use of lime, crucial for the intense red color, dates only to the sixteenth century," Isaac responded. "Before then some alternative natural plant or soil acid must have been used." He sighed, and then added: "But my investigations have uncovered no written or spoken records; the secret remains in a world we have lost."

While Isaac squeezed the limes, Maria prepared the cochineal insects for the dye. Since the supply in Mexico is now very limited, the weavers purchased their dried insects from Peru for about forty-seven dollars a pound, some 70,000 insects — a bargain since each bug must be tediously handpicked, swiftly killed by immersion in boiling water, which also dissolves its wax coating, and then dried in the sun for several weeks. As an emergency reserve, Isaac maintains a small colony of the insects on a grove of nopal cactus growing in his backyard.

Maria placed a *metate*, or "grinding stone," in front of her on the patio floor, and using a stone rolling pin, ground the dried bugs in much the same way she would grind corn for tortillas. Within thirty minutes, she had transformed one-half pound of the drab bugs into a colored powder. The grinding complete, Maria removed the powder and rinsed the *metate* with a cup of water, carefully collecting the runoff: Cochineal is too valuable to waste.

When the water was boiling, Isaac added the powder and the salvaged dregs, and the solution became a dark red. He stirred the dye for a few minutes with a stick bearing the tenacious stains of past dye pots, then added the lime juice. The dye turned a brilliant red. Finally the three skeins of wet woolen threads were lowered into the boiling cauldron, and the solution was kept simmering for one and one-half hours.

Soon I understood why Isaac Vasquez is an internationally esteemed master of his craft. The wool's final color depends on many factors, including its initial color, ranging from white to various shades of gray, brown and nearly black; the length of time the wool is in the dye bath; the amount of cochineal powder used; and the amount and type of lime used (Isaac used the juice from fresh limes, but his cousin Alberto prefers whole dried limes.) By manipulation of these factors, countless hues within the red spectrum can be created. Since the precise conditions are never recorded, however, only his years of experience and perceptive eye endow Isaac with a sensitive control of these natural colors, a control that can be admired and even envied, but never copied.

When Isaac gauged that the wool had reached the appropriate color, the skeins of thread were removed from the dye bath, shaken vigorously to dislodge the fragments of *tejute* leaves and limes, and hung on a nearby tree limb to cool and dry overnight. The exhausted dye bath was then thrown out. When Isaac needs a relatively muted pastel, he will reuse the solution immediately for dyeing additional wools. He may also use it for overdyeing, a process in which threads previously dyed with one color are redyed with a second to achieve an uncommon hue. Because there is no refrigeration to protect the organic ingredients from spoiling, at day's end the solution in any case has the same fate. It is poured onto the bare ground of the patio.

PHOTO BY AUTHOR

The reds in this hand-woven Mexican tapestry were all derived from cochineal.

The following day, we packed the pungent threads into my pickup and travelled the three miles of rocky road above the village to a small trickling stream. There Isaac washed the threads in a bucket of sudsy water (Fab is the product of choice) and rinsed them thoroughly in the flowing stream. The *tejute* leaves had done their job once again; the threads retained their intense red color. Back in the village, the skeins were suspended from a tree limb for two to three days of final drying before being transformed on the loom into works of art. ❖

A LONGER VERSION OF THIS ARTICLE FIRST APPEARED IN THE MARCH 1986 ISSUE OF *NATURAL HISTORY* MAGAZINE. REPRINTED WITH PERMISSION.

DYEING IN THE HIMALAYAS

Diana K. Myers

The Tibetan culture area of South Asia has as its axis the Himalayas, a spectacular range of peaks stretching from Afghanistan in the west to Burma in the east. Lying in the tropics, this area boasts some of the most dramatic topography in the world. Lush forests blanket the hills which rise steeply from the Indian plains to become the highest mountains in the world. North of the main range is the arid, windswept Tibetan plateau, lying over 12,000 feet high in the rainshadow of the Himalayas. The sphere of Tibetan cultural influence thus includes diverse climatic zones: Ladakh in northwest India and Tibet itself are part of the high-altitude plateau, while Nepal and Bhutan are nestled on the temperate slopes south of the permanent snows. This essay incorporates observations from Ladakh, Sikkim and West Bengal in India; Nepal, southern Tibet, and Bhutan.

The cultures of these distant areas were linked historically through trade and political alliances as well as through Buddhism, introduced to the region in the seventh century. It was not only religious pilgrims, diplomats and merchants who criss-crossed these long distances south of the famous Silk Route. Members of the central Tibetan royal family fled upheavals in Lhasa and founded a dynasty in Ladakh in the tenth century, while a remarkable abbot from southern Tibet united Bhutan as a Buddhist state in the seventeenth century. The cultures of Ladakh and Bhutan evolved quite distinctly from that of Tibet although the arts, crafts, costumes and other aspects of these far-flung communities show important similarities. Nepalese culture includes Tibetan Buddhist elements as well, although today the kingdom is Hindu.

Early Western visitors to the Himalayas recorded little information on weaving and dyeing, so much of our knowledge of traditional practice is inferred from present custom. We do know that wool, salt and borax from Tibet and Ladakh were exchanged across mountain passes for cotton, silk, paper, medicinal herbs and dyeplants from the south. The impact on this trade — and therefore on dyeing materials and methods — of twentieth century political events, which first brought an influx of manufactured goods and then closed Himalayan borders, cannot be underestimated. Nevertheless, some communities have never abandoned or are now re-adopting the weaving materials and dyeing practices of earlier generations.

The arts of weaving and dyeing are widely practiced throughout the Hi-

---

Diana K. Myers *began studying Himalayan weaving while serving in the Peace Corps in Nepal. She is the author of* ***Temple, Household, Horseback: Rugs of the Tibetan Plateau*** *(The Textile Museum, 1984) and now directs development and training programs for the Experiment in International Living.*

malayas, especially in rural areas where the change is slow to occur. Fabrics are woven for a tremendous variety of utilitarian goods: costumes, animal trappings, blankets, pile rugs, tents, doorway and floor coverings, cushion covers, bags and numerous ceremonial textiles used in monasteries and in secular and religious ritual.

The most common weaving fibers are sheep's wool, cotton and raw silk. Yarns may be left natural or tinted, and dyeing is generally done prior to weaving. In some Himalayan areas dyeing is a specialized occupation and in others it is carried out more widely. This appears to be partly a function of geography and climate, for dyeing requires large amounts of fuel (wood or dung) to heat the dye vats, sometimes for days. Thus it is done less in barren regions of Tibet and Ladakh than in temperate valleys where forests and animal herds provided plentiful fuel. In Ladakh, dyeing for each community is done by only a few households, because dung is limited and wood expensive. In heavily forested Bhutan, dyeing is done in many more homes and generally by women. Some individuals are known as particularly skilled dyers, but virtually every woman who weaves knows how to color her yarns.

Himalayan villagers traditionally employed a standard selection of dyeplants obtained through trade and supplemented those with local materials or variants. The most important of these survive in use today, both as sole coloring agents and as overdyes to enhance the color of commercial yarns. Synthetic dyes arrived gradually in the Himalayas and their impact was uneven. They were available in some areas as early as the 1870s, but other regions did not see their advent until the 1950s. Nowadays, commercial dyes, yarns and fabrics predominate, because they are inexpensive, easily available and labor-efficient. At the same time, traditional textiles, dyeing materials and techniques are valued. Many rural Bhutanese, for example, are fond of the vibrant colors resulting from store-bought dyes, but consider synthetic tints inferior to hues achieved with natural dyes. Bhutanese in urban areas are leading a change in fashion where subtle colors and traditional design are once more favored over polyester and brightly tinted garments from abroad.

The most important traditional vegetable dyestuffs are madder, indigo, rhubarb, myrobalans, sweetleaf, walnut, lac and wood sorrel. Historically, safflower was also a major dye. Methods of extracting color from these substances are quite similar across the region, although recipes vary according to the materials locally available and the fiber being tinted. Mordants include natural salts and sulphates found on the banks of rivers and lakes and near hot springs: primarily borax, common soda, alum and copper sulphate. Millet or barley beer often serves as a reducing agent in indigo recipes, while rhubarb root and various fruits and leaves provide acidic juices for modifying and fixing other colors.

Madder
RUBIA SPP.

The most popular natural source of red, madder is the common name of several species of the *Rubia* plant, a creeper native to the Himalayas. Varieties are found in Ladakh and southern Tibet, but the hills of eastern Nepal and Bhutan are most abundant in the plant, which people would "gather in the jungle, cut...up into small pieces and carry...away into Tibet, where a rich red dye is extracted from it." Early visitors noted the important trade where Tibetan salt was exchanged for madder from Bhutan and Nepal; taxes were also paid with the dyeplant.

A Tibetan refugee in Nepal dyes wool with churtsa, rhubarb root.

South of the Himalayas, wild madder stems and roots are harvested in April-May and may be used for dyeing when fresh or dried; if dried, they are sometimes ground into powder. Dyeing involved heating large quantities of madder with the yarn, which is treated with a mordant such as alum, myrobalan or soda before and/or afterwards. Shades obtained are pink, red and brown as well as orange (in combination with turmeric).

Indigo

STROBILANTHES AND OTHER PLANTS CONTAINING INDIGOTIN

For blues, Tibetans and other Himalayans relied heavily on indigo imported from China and India, where its commercial production had been of worldwide importance since the end of the eighteenth century. Dyers also obtained blue from local plants containing indigotin that were available in areas

Symplocos sp. contain a yellow pigment popular in the Himalayas as a dye and mordant.

near Burma and China. In Bhutan, blue comes from several of these broad-leafed shrubs collectively known as "indigo" (*Strobilanthes* sp.).

Leaves of these plants are gathered in the wild, dried and ground. The intricacies begin in processing the dye: First the cakes of powder are fermented in a solution of hardwood ashes, or cow's urine and water, or oxalic acid from rhubarb juice, or local beer, or yeast. Recipes differ, but all call for careful monitoring of this fermentation and the temperature level during successive steps in the dyeing process. Adjustments may be made in the recipe in order to achieve the desired color: The first dyelot is a light blue, with two to four steepings yielding deeper shades. The entire procedure may take up to six months in some areas, with attention paid to the season, the placement of the clay pots in which the dye ferments, and the number of times the mixture is stirred.

Not surprisingly, the indigo-dyeing process is easily spoiled. Hence Tibetans believe it should be done by one person (specifically not a menstruating woman), in a quiet place where no one can observe the first steeping of the yarn. Bhutanese feel pregnant women should not be involved lest the color be "stolen" by the baby; nor should strangers witness the activity, lest they take away the color.

Rhubarb
***RHEUM* SPP.**

The most common source of yellow in Ladakh and parts of Tibet, and today among Tibetan refugees in Nepal, is rhubarb root. Rhubarb species, which abound from one end of the Himalayas to the other, have long been sought after as a local dye. The Mongols likewise valued the roots for coloring. In some areas, the roots served as medicine, too,

PHOTO BY AUTHOR

The woman spinning wool above is wearing a robe dyed with madder.

Shawl of white cotton patterned with raw silk dyed with indigo and madder.

and the leaves of the plants were smoked like tobacco.

Dug up in autumn, rhubarb roots are cut into pieces an inch or so long, dried and powdered. When boiled in a plain or salt solution with yarn, the dye yields strong, bright shades of orange, gold or yellow. With soda as a mordant, yarn dried in the sun turns pink. Less commonly used, rhubarb leaves tint yarn pale yellow.

Myrobalans

***TERMINALIA CHEBULA, T.* SPP.**

Along with rhubarb root, myrobalans are common agents for yellow. Sour fruits of trees of the *Terminalia* genus, which grow below 5,000 feet, myrobalans are mentioned as a major Himalayan dye as early as the mid-seventeenth century. Important, like rhubarb, for medicinal purposes (the Buddha of medicine holds a myrobalan leaf), they yield a golden brown color or serve to darken indigo and other dyes. The two most common species (*T. chebula* and *T. bellirica*) have a high tannic acid content which makes them excellent mordants as well.

The fruits and galls of the leaves, mixed with alum, produce "a most durable yellow" on cotton, silk and wool. Mixed with iron salts, the fruits turn wool a dark gray. With safflower, madder and other dyestuffs, the astringent fruits function as a mordant.

Sweetleaf

***SYMPLOCOS RACEMOSA, S.* SPP.**

The leaves of at least three species of *Symplocos* contain a yellow pigment and are popular as a Himalayan dye and mordant. *Zhungkhe* (*S. racemosa*) is one of several related shrubs native to the higher hills of India, Sikkim and Bhutan. The leaves, and sometimes the bark, are gathered in the wild (in the Darjeeling area, they are picked in spring) and dried for dyeing.

Leaves are chopped before being boiled with cotton or wool. Very often turmeric and coarsely ground buckwheat grains may be added to the dyebath to heighten color, while madder leaves and twigs are stirred in to produce orange and rust. *Symplocos* is also an important mordant used with madder and with local sources of indigo.

Walnut

JUGLANS REGIA

Walnut trees grow throughout temperate regions of the Tibetan plateau and southern Himalayan hills. The husks of their fruit provide the main brown dye and serve as a darkening agent. There are several methods of using walnut as a dye. In Ladakh, green husks steeped in boiling water yield a light yellow-brown, fixed with salt. Alternatively, dried husks may be boiled whole to achieve browns, primarily on wool yarn. Elsewhere the husks of unripe walnuts are stored in pots of water to rot before being dried and powdered; various shades of brown result from this dyebath, whose acidity is varied by adding rhubarb juice or soda. In the eastern Himalayas, informants describe using the bark or roots of walnut for paler tints of brown, but say these materials do not produce the same strong colors.

Lac

A SECRETION OF THE *LACCIFER LACCA* INSECT

The only regional dye cultivated as an industry of sorts is lac. Like madder, it was an important Bhutanese export to Tibet, Nepal and India, and to a lesser extent to Ladakh. While lac production was never very systematic and its scale has greatly diminished in this century, in the warm eastern valleys of Bhutan, landowners still register the trees on their land which are used for rearing the lac-secreting insects.

From May onwards, these insects feed on several shrubs, the most important of which is *Butea monosperma*. By fall, they have secreted hard "cocoons" which attach them to the twigs; at this time the twigs are gathered and hung on the branches of a type of shrub (*Zizyphus* sp.) where the insects continue to feed and lay eggs over the winter. When the weather warms, the shell-like encrustations, known as stick lac, are harvested and the next generation of insects is moved to the *B. monosperma* once again.

For dyeing, the lac is crushed and then made into a paste with water and yeast. Wheat flour is added and the mixture fermented in a clay or stone pot. In Sikkim, this preparation is warmed over a low fire for one week and in Ladakh lac dyeing is done in the summer when the sun hastens fermentation. During this gradual heating, a white residue rises to the surface and is skimmed off. Wool is then soaked for up to 20 days in the solution. In Bhutan, the solution is boiled down to obtain a dye concentrate, and fibers are prepared for receiving the color by steeping in a solution of sweetleaf.

Various ingredients are added to modify the shade. With alum, lac produces crimson; with iron, purple; and with salt, scarlet. Maroon is obtained by immersing lac-dyed yarns into a solution of indigo.

Wood Sorrel

RUMEX NEPALENSIS

Greens are usually achieved by overdyeing a blue yarn with rhubarb or barberry, but several plants produce greenish hues. The best color is considered to come from *Rumex nepalensis*, known in the Darjeeling area as "dog leaf." A shrub found from Bhutan to Kashmir and in the Chumbi Valley of Tibet, its leaves yield a light green which is darkened by adding the roots to the dyebath. This variety of *Rumex* should not be confused with common sorrel (*R. acetosa*), whose roots produce red, or with common dock (*R. obtusifolia*) whose roots produce black; neither seems to be known in the Himalayas.

Safflower

CARTHAMUS TINCTORIA

Another important source of red, especially for woolen fabrics, is Tibetan safflower or wild saffron, distinct from Kashmiri saffron, which produces yellow rather than red. Native to temperate areas from Bengal to western China, but not found in Tibet, this plant grows about three feet high. Its stamens contain the pigment carthamine. Picked in spring, the flowers were exported to Tibet and used to color narrow lengths of woolen twill then traded to China and other regions. Because of its association with this famous fabric, the distinctive color safflower produces was known to the Chinese as "Tibetan red."

Recipes for dyeing with safflower call for the addition of soda or myrobalans. It was also used as a compound tint with indigo. Outside Tibet, it was mainly used in combination with lemon juice or other acidic mordants to dye cotton and silk.

Today in the Himalayas, indigo, madder, rhubarb root and walnut are the most prevalent natural dyes, but lac and myrobalans are also popular. In Sikkim and Darjeeling, *Symplocos, Rumex nepalensis* and barberry are widely used. Some of the less familiar plants are collected as they always have been by villagers residing in high-altitude valleys from Ladakh to Bhutan. The colors obtained, however, vary with the chemical content of local water as well as with the mordants used — and still depend greatly on the dyer's skill. ❖

For Further Reading:

Sasuke Nakao and Keiji Nishioka. *Flowers of Bhutan*. Asahi Shimbun Publishing Co., 1984.

KALAMKARI

Dye-Painted Fabrics of Sri Kalahasti

M. Joan Lintault

Southeastern India has long been famous for its richly decorated, naturally dyed cotton fabrics. They have been known throughout Asia for more than 1500 years, and traded to the West since the 16th century. From the time of their introduction, the exotic patterned fabrics called "chintz" (from a Sanskrit word meaning spotted or bright) were all the rage in Europe. Indian cotton cloth was valued not only for the beauty of the designs, but also for the richness of its colors and its wearability. There was nothing equal to it in the West. Chintz was used for garments, bed clothes and home furnishings.

In India, bright-colored decorated fabrics are called kalamkari. These can be either block printed or hand painted. A special type of kalamkari comes from the town of Sri Kalahasti in the state of Andhra Pradesh. These fabrics feature handpainted designs that depict stories from the great Indian epics. Traditionally, they might have been used as narrative picture scrolls by storytellers who recited tales to an audience, or displayed on temple walls during festivals. Now the kalamkari painting process is used to decorate cloth for wall hangings, door curtains, table cloths, saris and other articles.

The art of making kalamkari had nearly died out in the 1950s when the All-India Handicrafts Board engaged the last surviving kalamkari expert to start a school and teach the craft to five young boys, including his son Gurappa Chetty. I visited Sri Kalahasti, and met Gurappa Chetty. He is very much a traditional craftsman, and inherited from his father a reverence for the Kalamkari patterns that have been handed down for generations. Although each Kalamkari artist has his own style of line quality and

Guarappa Chetty applies alum mordant.

A closeup of the artist drawing freehand.

M. JOAN LINTAULT *teaches art at Southern Illinois University in Carbondale, Illinois.*

graphic expression, there is a similarity in the colors, the divisions of the compositions, and the forms, postures and expressions of the figures.

Kalamkari artists never draw from life or nature. Their subjects are inspired by the words of the epics. They paint scenes across the fabric from left to right, accompanied by commentary written in Telegu, the language of Sri Kalahasti. Some artists outline their drawings on the cloth with charcoal made from tamarind twigs, but Gurappa Chetty usually paints freehand without benefit of preliminary sketches. He is able to maintain correct proportions of the body, and has a fluid style of line and an amazing sense of pattern and decoration.

Although Gurappa Chetty devotes all his time to his art, he does not have a studio. All his work is done on his front verandah. In India, life happens in the open, not behind closed doors. Everyone speaks to him as they pass by. Vendors, children and friends visit. Animals slowly wander down the street. Life swirls around Gurappa Chetty as he paints his kalamkari.

Kalamkari Technique

Making a kalamkari requires a series of steps that extend over a period of several weeks. First the fabric, a cotton muslin called gadda, is washed well to remove all the starch and sizing. Then it is mordanted with a solution of powdered myrobalan fruit (*Terminalia chebula*), which is high in tannic acid, and water buffalo milk. The milk is important because its high fat content coats the fabric and increases dye absorption, so the dyes then may be applied without wicking. The cloth is soaked in the solution, then tightly wrung several times, and spread out in the sun to dry. It has a yellowish color at this stage.

A kalamkari artist uses a homemade pen called a kalam, made from a piece of bamboo or other straight stick sharpened to a point at one end and wrapped with a rough piece of cloth or hair held in place with string. The artist dips the kalam into dye, which soaks into the cloth, then gently squeezes the cloth to make the dye run down the stick and onto the fabric. It takes a great deal of skill to use just the right amount of pressure to avoid making blots on the design.

The first dye to be applied is called kasam. It is a clear liquid made by soaking pieces of iron in palm sugar syrup (jaggery) for 21 days. This is the traditional method for making iron acetate. When this dye is applied to the myrobalan-mordanted cloth, the high tannic acid content of the myrobalan combines with the iron and gradually turns the cloth a permanent jet black. This dye is used to create the outlines of the design.

After the outlines have been drawn, the cloth is prepared for red dyeing. A kalam is used to apply a solution of alum water to the areas that will be red, pink, orange or purple. The cloth is dried, then rinsed very well in swift-flowing water. This is an important step, as careless rinsing results in the alum spreading to other areas.

Then the alum-mordanted cloth is dyed in a solution of pobbaku leaf (*Nagrigama alta*), surudu root bark (*Ventilago madraspatana*) and majnistry or madder root (*Rubia cordifolia*). The powdered dyestuffs are mixed with water, the cloth added, and everything is boiled for one hour in a brass, copper or pottery vessel. When removed from the dyepot, the alum-treated areas are darker red, and the surrounding areas are pink. If several shades of red are required, the fabric is dipped again in the myrobalan/milk solution, then treated with alum and redyed. Repeated dyeing gives a deeper, more permanent red.

At this point, the cloth has an overall tinge of red, which must be removed so that the other colors can be applied to a clear background. To bleach out the unwanted red, the cloth is soaked overnight in sheep or buffalo dung mixed with water, then squeezed and spread on the wet sand of a riverbank. For several days to a week, it is sprinkled repeatedly with water and turned so that both sides are exposed to the sun. There is some ingredient (perhaps ammonia or uric acid) in the dung of a vegetable-eating animal that, combined with the sun, has a bleaching effect. After this treatment, the portions of cloth painted with iron and alum solutions retain their colors (black and red, respectively) but the rest is bleached white.

The next step is dyeing with yellow. The cloth is soaked in buffalo milk, and wherever yellow or green colors are desired, dye is applied with the kalam. The yellow dye is made from powdered myrobalan, mango bark or pomegranate husks, boiled with water and alum. The dyed cloth is dried in the sun, then washed in running water.

Blue is the last color to be added. Traditionally, wax was applied to all the areas that were not to be dyed blue. Then the cloth was dyed in indigo. Indigo over yellow gave green. Today most kalamkari craftsmen use a synthetic blue dye instead of indigo, and apply it with the kalam. Then the cloth is washed and dried one final time.

GROWING MADDER

Rita Buchanan

Madder (*Rubia tinctorum*) is a hardy perennial native to the Middle East, with slim four-sided stems, whorls of slender pointed leaves, and tiny white flowers. Weedy-looking and prickly-textured, it sprawls over the ground to form a low wide mat. Madder seeds are rarely listed for sale, but you can order plants from several mail-order nurseries. Buy one or two to start with, and you can divide them over time to multiply your stock. Prepare a madder bed by digging at least 12 inches deep and working plenty of compost, manure and lime into the soil. In late spring, set plants 18 inches to 24 inches apart. When the madder tops get a foot tall, pull them down to the ground and draw a little soil over the middle of the stems so that they will form new roots. Do this several times a season for three years in order to make a solid bed of plants and roots. In spring of the third year, pull out a few plants to start a new bed in fresh soil. That fall, use a digging fork and work through the soil to harvest your crop of madder roots. There will be quite a tangle of them, thick as pencils and bright red-orange. Wash the roots thoroughly, and use them fresh or dry them for storage.

These women are waiting for the dyed cloth to dry before dipping it in the river again and again to refine the color.

ENDURING DYE TRADITIONS

OF CHINA'S MIAO AND DONG PEOPLE

GAIL ROSSI

Ancient dyeing methods still play an important part in the everyday lives of women inhabiting remote mountain regions of China's southwest in Guizhou, Guangxi and Hunan provinces. The geographical isolation of these rugged areas, where nationalities such as the Miao and Dong live, has helped preserve a wide range of unusual textile traditions. Making and wearing unique traditional costumes not only maintains cultural identity, but also remains an important way for a young woman to attract a desirable marriage partner. A woman's festive garment is a reflection of her creativity, diligence and skill.

GAIL ROSSI *is a weaver and embroiderer. She has taught English in Beijing and has studied the textiles of China for several years.*

PHOTO BY AUTHOR

Lengths of indigo-dyed fabric dries on rafters in Dong village.

Although synthetic dyes are used in these areas, many women prefer the lasting beauty of the time-honored natural dyeing methods, which use such materials as indigo, oxhide, beans, barks, pig's blood, leaves, egg whites and even red peppers to produce special effects. Requiring sometimes over one hundred steps, cloth is dyed, dried, redyed, beaten, smoked, and steamed in processes which differ from nationality to nationality, and from village to village within the same nationality.

When living in China for nine years, I was able to visit the province of Guizhou over thirty times to document the area's rich display of traditional textile art. I visited the Miao district of Zhouxi many times and lived with a village dyer's family. This family has been in the profession since the 1950s. It carries a certain amount of prestige in the area, and people come from neighboring villages to deliver undyed bolts of handloomed cotton cloth. According to their requests, it is either dyed chemically (a quicker and cheaper process) or dyed in the ancient way. Many women prefer the traditional method, since the results are especially beautiful and never fade.

Traditional dyeing in Zhouxi includes many steps. Between all dippings, from beginning to end, the cloth must be thoroughly dried. First, the white cloth is dipped three times in natural indigo. Then it is immersed eight times in a white liquid made from pressed beans, which seems to help the cloth absorb color. The fabric is again dipped in indigo, three times a day for five days.

Next it is dipped about seven times in a reddish-black liquid made by boiling a wild tree root in water. This blackens the cloth. It is then placed two or three times in a watery-gluey substance made by boiling an ox skin in water. The hide liquid stiffens the cloth and makes it colorfast.

The fabric is dipped once in pig's blood, which adds a reddish sheen. It is then placed in a wooden bucket and steamed for one hour. This helps make the dye adhere. The fabric is then beaten with a wooden mallet on a large flat stone. Finally it is dipped in indigo again, until the desired effect is achieved.

In this area, it's a common sight to see groups of women down by the stony river bank, dipping their cloth in buckets of dye and then laying long lengths on pebbles to dry under the sun. While the professional dyer is constantly busy dyeing others' cloth, some individuals continue to dye their own. Most households have a huge indigo dye pot for simple indigo dippings.

The dyed fabric is made into festive garments which are further decorated with traditional silk embroidery, silk felt appliqué, fine tin embellishments and decorative weavings. Most families raise silkworms, and the women spin fine silk threads for both embroidery and weaving. Commercial dyes from the market are used to dye the threads in bright colors, but a brilliant yellow is obtained with seeds from a small orangish-yellow pod from a local tree. Put into cold water, the seeds instantly produce a vivid yellow dye which readily adheres to the silk threads. The color lasts for many years, even after repeated washings.

Professional dyers in Zhouxi buy indigo paste from professional indigo growers who live in the Wanchao district, about 50 miles east. Though three kinds of indigo grow wild in the Zhouxi area, the dyers prefer the dependability of the purchased indigo. Over three hundred indigo-producing families live scattered throughout the mountains of Wanchao. Large stone pools hewn from the mountainsides are a common sight there; each pool is capable of holding up to a thousand soaking indigo plants. Indigo paste is sold in local markets and ordered by professional dyers who covet its rich blue color. Paste from the annual harvest is often stored in leaf-lined baskets, covered with a layer of mud and submerged in fresh spring water.

A Miao group living in Huangping and Kaili counties of Guizhou use a form of what may be gentian violet to color their unique traditional fabric. They spread the dye onto the fabric with a brush, then hold it over a smoldering fire of smoking cypress leaves. Then the cloth is beaten until completely dry. This process is repeated until the cloth turns a brilliant metallic gold. Such fabric is made into festive jackets and finely pleated skirts. The color comes off easily when touched—handling it for several minutes turns your fingers a bright purple-blue.

The starchy root of a terrestrial orchid known locally in Huangping and Kaili as "bing jiu" (*Bletilla striata*) is boiled in water to make a thick liquid used to coat fabrics for several purposes. Some women spread the liquid on the underside of their patterned weavings, to help strengthen the cloth and to keep the threads from unravelling. Likewise, the liquid is brushed onto pleated skirts to help set the pleats.

The Dong nationality live further south, in southeast Guizhou extending down to Guangxi and west to Hunan. In Dong villages, the air vibrates with the constant sound of beating — women beating the warps on their looms, and women beating their finished dyed cloth to perfection. Everywhere lengths of brilliantly dyed cloth hang to dry from carved wooden porches. The dyeing process is basically similar to what has already been described, using combinations of barks, indigo, pig's blood and leaves, but the Dong include a further ingredient — egg whites.

I visited two women who had spent an entire day pounding long lengths of dark reddish-brown cloth. That night they would spread beaten egg whites onto the fabric, then steam it for several hours. The resultant cloth shines like glass. Egg-white application is used by a number of Dong dress groups to achieve quite different results. Some Dong apply egg whites to the tight folds of finely pleated skirts. When steamed and dried, the pleating stitches are let loose. The effect is like shining stars; the egg resting on the outer folds reflects light, especially as the wearer moves. In the same area, several Miao groups also employ the egg white method. One particular Miao group mixes egg whites and ground red peppers. When applied to the cloth and steamed, the resultant sheen is a silvery purple-red.

Another group of Miao in this region dress much like the neighboring Dong. What differentiates the two ethnic costumes is a dye-resist decorated cloth made by the Miao. Unlike the wax resist used in other areas of Guizhou, the Miao here use a combination of pig's fat and the sap from a maple tree. This concoction is used to draw traditional motifs onto leggings, apron yokes, jackets and children's school bags. After dyeing in indigo, the substance is boiled off and the unique designs appear.

These ancient modes of dyeing are very much alive today in remote mountain valleys and highlands of southwest China. The Miao and Dong people continue to make and wear traditionally dyed garments. But ethnic groups near large cities favor easier-to-care-for Western clothing and fabrics. Such fabrics today make their way even to remote markets. It's not uncommon to see everyday garments made from polyester blends, though cut in traditional design. Nonetheless, women still prefer to use traditionally dyed fabric for festive costumes. They consider it most precious and beautiful — the perfect backdrop for their exquisite embroideries, weavings and abundant silver jewelry. ✣

The Vegetable-Dyed, Mud-Mordanted Silk of Kumejima

Dorothy Miller

Walking down the streets of Kyoto, Japan, during the Gion Festival in 1957, I admired the women dressed in their kimonos for festival days. Especially appealing were the silk kimonos designed with splashed patterns of color, some designs being birdlike, and others geometrical.

One kimono textile of particular interest to me came from the weavers of Kumejima, an island in the East China Sea near Okinawa. It was dark brown in color with patterns of gold and orange, and was woven from handspun silk. I later saw similar fabric in an exhibition of Okinawan textiles at the Folk Craft Museum in Tokyo, and I sought more information on the dyeing process. I learned that the rich colors are achieved by dyeing the fibers many times in vegetable dyes and mordanting them in a mud bath. The process is called *doro-zome*, which means "mud-dyeing" in Japanese. Finally in 1975 I was able to go to Kumejima to see how this fabric is made.

History

Long ago, the weaving and dyeing of silk was not done solely for the people of Kumejima or for those of the nearby island of Okinawa. Instead, silk textiles were woven to pay the taxes exacted by the feudal lords of the Satsuma clan who governed the islands. The story goes that women were chained to their looms to weave enough cloth to meet the demands of the court. The Satsuma regime controlled their subjects until the early 1900s, when Japan was reorganized under an emperor. When they were freed from feudal taxation, the Kumejima weavers stopped their work and the traditional art began to die out.

Years later, however, the weavers became aware that they could make money by selling their goods in the mainland cities of Kyoto and Tokyo, and cooperatives were formed for marketing Kumejima silk textiles. The weavers went back to work raising silkworms, dyeing fibers and weaving cloth. Interest in these textiles was furthered when Shoji Hamada and Soetsu Yanagi visited Okinawa in 1939. Mr. Yanagi wrote about the crafts of the islands in that part of Japan and Mr. Hamada promoted interest in the crafts through the Folk Art Society of Japan.

Today the making of Kumejima silks can be seen at a weaving studio called Kumeai, where islanders are taught the process. I visited the studio, and also saw weavers at their looms in the village near the Kumeai. On the beachfront road nearby are modern hotels put up for the Japanese tourists.

Dorothy Miller *has been researching textiles of the Far East for over 30 years. One of her specialties is indigo dyeing, and she is author of the book* ***Indigo from Seed to Dye*** *(Aptos, CA, Indigo Press: 1981).*

Weavers sell their fabrics through a co-operative, and individual weavers also sell to the buyers from department stores and kimono shops in the cities of Japan. It is hoped that the weavers and dyers will continue to produce these beautiful kimono fabrics. Much depends on whether the younger generation will follow the ways of the older women of the island.

PHOTO BY AUTHOR

Silk fibers dry after being overdyed in a kuru dyebath and mud mordanted.

How the Fabrics are Made

The silk used in these kimono fabrics is spun from the outer fibers of waste cocoons. These fibers are removed in a sheet from the cocoon and spread over a type of distaff made for this purpose. They are hand spun into yarn called *tsumugi,* Japanese for "handspun silk," which resembles the tussah silk woven in India. The spun yarn is made into skeins for dyeing.

Dyeing is done at the time of year when the barks and roots are gathered. Plants used for the dark brown include *tikachi* (*Raphiolepis umbellata*), *kuru* (*Dioscorea rhipogonoides*), and *muwagi* (*Myrica rubra*). The yellow comes from *fukuji* (*Garcinia* sp.) and *kurobo* (*Diospyros maritima*). The red and rust colors come from *hachimachiban* (*Carthamus tinctorius*). The dyestuffs are boiled in an iron pot, then the dye liquid is poured off into a separate container and the silk yarns are soaked in this bath. The yarn is dyed repeatedly, as many as 18 to 35 times, in different dye baths, beginning with the lighter-colored dyestuffs and proceeding to the darker dyes.

The mud used for mordanting is gathered from mountain slopes where the soil is particularly rich in iron and manganese. In the past, weavers would place the yarns in natural mud baths, but nowadays the mud is collected and brought to barrels at the dyeing site. Some women wrap their yarns in mud-soaked canvaslike cloths. Others bury the yarns in barrels of mud for two or three days at a time. After the mud treatment, the yarns are rinsed and dyed again until the desired rich smoky brown color is achieved.

To make the splash pattern effect in the woven fabric, some of the yarn is dyed with a technique called ikat dyeing. To do this, yarn is measured to the length and width of the fabric to be made. Then these measured skeins are wrapped tightly — traditionally with raffia-like straw, but nowadays with plastic tape — wherever a pattern is desired. The wrapping keeps the dye from penetrating, and the wrapped areas end up lighter colored than the rest of the yarn. After the dyeing is completed, the wrappings are removed and the yarn is ready for weaving. The weavers create both warp and weft ikat patterns, and also make striped and checked designs. ❖

Further Reading

Hisao Suzuki and Tsune Sugimura. *Living Crafts of Okinawa.* New York, Tokyo: Weatherhill, 1973. (Text in English.)

Soetsu Yanagi. *The Unknown Craftsman.* Tokyo: Kodansha, Int., 1972. (Text in English.)

NATURAL DYEING IN TURKEY

JENNIE WOOD

A powerful resurgence of interest in natural dyeing began in Turkey during the 1980s, even though the traditional dye recipes and procedures had been abandoned and mostly lost when chemical dyes were introduced at the turn of the century. An important reason for this renewed interest is that Turkey's many-centuries-old tradition of rug weaving never died, and a significant portion of the market is showing its preference for and willingness to pay for rugs in the old colors of plant-dyed wool yarns.

This dyeing renaissance did not begin at the grassroots level with the weavers themselves. Rather, it was initiated by rug collectors and scholars, and is being promoted by several rug dealers who organize the production of oriental carpets and flat-woven kilims. For example, George Jevremovic, owner of Woven Legends in Philadelphia, is a dealer who sells fine old rugs, but he also supports contemporary weavers in eastern Turkey who make pile carpets in traditional designs, using handspun, naturally dyed yarn. In this context, George asked me to go to Turkey in 1988 to help the dyers expand the quantity and improve the quality of the colors they were getting. I visited dyers in different settings and will describe two of these: Malatya in the east and Ayvacik in the west.

Eastern Turkey is still rather wild and frontierlike, with long distances between towns, steep snow-capped mountains looming over the high plateaus, and vast expanses of dry rocky terrain. Malatya, a city of 250,000 in the southern part of this region, is an agricultural center, the apricot capital of Turkey. There I spent a week working at a dyehouse supervised by Mehmet, a master dyer, whose brother works for George in Istanbul.

The dyeshop is a basement area, approximately 30 feet square, in a building that serves as a warehouse for the carded wool that is supplied to local handspinners and a collection and distribution center for the finished yarns. Three tin-lined copper kettles, each big enough to hold thirty kilos of yarn, sit over gas burners fueled by propane tanks. The kettles are filled with a hose and emptied through valves at the bottom.

I noticed that yarn is never washed before going into the kettles. Skeins are simply put on the floor and hosed down with cold water, then immersed into the hot baths for mordanting or dyeing. When skeins are taken from the dye kettle they are put on the floor and again hosed down with cold water. The sudden

---

JENNIE WOOD *has been studying and practicing natural dyeing for 15 years, with a special interest in ethnic dyes. She works as manager of a wool-spinning mill in Philadelphia, Pennsylvania.*

temperature changes result in some felting of the wool, making the bundles of yarn a bit difficult to use, but no one seems to consider this a serious problem.

For most colors, yarn is pre-mordanted with alum at the rate of 15% of the dry weight of the wool (3 kilos of alum for 20 kilos of yarn), and cream of tartar at 5%. This combination is the only mordant I saw used, but the dyers had some iron on hand and were familiar with its use in producing browns and blacks. Mehmet tries to maintain a supply of alum-mordanted yarn ready for the dyepot. Dyeing is usually done by the simultaneous method: Plant materials and yarns are simmered in the kettle together.

In the southwestern region of Turkey, on the Aegean coast, I visited Ayvacik, a prosperous-looking market town of 10,000 that serves as a commercial center for a number of weaving villages in the area. Several streets in town are lined with small shops that sell yarns, carded wool for spinning, bags of chopped madder root, mordants and kits for indigo dyeing. On Fridays, weavers come to town to sell their carpets to rug dealers, tourists and Turkish retail buyers.

In Ayvacik I worked for several days with Ahmet Balci, a home-based dyer. Ahmet uses two tin-lined copper kettles, each holding approximately five kilos of wool, to dye yarns for his wife and for other weavers in the area. He works outdoors over wood fires, but uses many of the same raw materials and procedures as the more commercial operation in Malatya.

Colors and Dyes

Dyeing for the oriental rug palette concentrates on the three primary colors: yellow, red and blue. There are many plant sources of yellow in Turkey, as there are in most places; the plant chosen for use depends in part on regional differences in flora. On the other hand, almost all the red dyes come from one plant, and the blues all come from indigo.

YELLOW. I saw three plants used for yellow. First was a spurge, probably *Euphorbia biglandulosa* or *E. peplus.* Its Turkish name translates as "milky," and I can attest to its having a potent milky sap — I picked some along the roadside, later rubbed my eyes and had to go to an emergency room for treatment! In Eastern Turkey, the villages around Malatya gathered the spurge plant and dried it for delivery to the dyehouse. The dyers filled the kettle in layers, adding 15 kilos of alum-mordanted yarn, then a kilo of dried spurge, then more yarn and more spurge. They brought the dyebath to a simmer and cooked it for an hour, obtaining a yellow with a faint greenish cast.

The Malatya dyers used a similar layering procedure with onion skins to produce a slightly warmer and rosier yellow. In Ayvacik, Ahmet used dried blossoms and stalks of some kind of chamomile, probably *Anthemis chia* or *A. tinctoria,* to dye yellow.

RED. It's hard to exaggerate the importance of madder root as a source of reds for the Turkish dyer. The plant, *Rubia tinctorum,* grows wild in Turkey; it has not been cultivated for many years. The Turkish name for madder translates as "root dye." The roots are thoroughly dried, then chopped into small pieces or preferably ground into a powder.

In Malatya, the dyers used coarsely ground madder at the ratio of 25% to 50% of the weight of the yarn. Typically, the madder was not pre-soaked, but was put directly into the dye kettle and simmered for about two hours. Yarn premordanted with 20% alum and 10% tartar was added and simmered for an hour or more, resulting in a dark brick red. Then a batch of unmordanted yarn was added to the same kettle, along with 15% alum and 1% tartar, and simmered for an hour, giving a red-orange. The process was repeated until the madder bath was exhausted, producing red-orange, yellow-orange and finally two tints of a lovely peach color. For a lovely clear pink, the Malatya dyers dipped unmordanted skeins into a madder dyebath for less than a minute, followed with a dip in a hot alum bath before the final rinsing.

Madder can be used in combination with other plant dyes. Ahmet used it with chamomile to get rosy yellows and salmons, and Mehmet used it with spurge. Combined with walnut hulls and a tannin source like oak galls or acorns, madder gives a variety of warm reddish browns. And madder is occasionally overdyed with indigo to produce purple.

Blue. The dyers I visited in both eastern and western Turkey used crystalline synthetic indigo, which is chemically identical to natural indigo but much less expensive and more readily available. The Malatya dyers begin by dipping the skeins to be dyed in a hot alum bath. Then they heat the dye kettle to 95 degrees to 105 degrees F and add four ingredients: 1)"Tuxtal," an extract of boiled bones which thickens the bath and helps keep the indigo from settling to the bottom; 2) Sodium hydroxide or lye, which makes the bath alkaline; 3) Sodium hydrosulfite, which removes oxygen from the water; and 4) Indigo. The dyers stir the solution to dissolve the ingredients, wait ten minutes, then add about 30 kilos of yarn, nearly filling the kettle. After nine minutes, the top layers of yarn are removed and hung up to drip. Six minutes later, the second batch is removed, followed fifteen minutes later by the bottom layer, the darkest blue. All the skeins gradually turn bluer as they hang in the air. When the colors seem stable, the yarn is hosed down on the floor with cold water. The result is a wide range of blues with considerable "abrash" or streakiness — an effect that many rug collectors like to see, because it makes a textile seem older and more tribal.

In Ayvacik, Ahmet uses the same ingredients and in approximately the same proportions as in Malatya, but in much smaller quantities. He doesn't start by dipping the yarn in an alum solution, and he puts completely dry yarn into the indigo bath. He leaves skeins in the dyebath for three to five minutes, takes them out to air and dips them again if they are not dark enough. He may repeat the dipping and airing several times, then finally hoses the skeins with cold water. I was surprised that the temperature of his indigo bath was 150 degrees F — so hot that it shouldn't have worked! But Ahmet's indigo skeins have much less "abrash" than those from Malatya, and less tendency for the color to rub off.

Conclusion

As I traveled around Turkey watching both dyers and rug weavers at work, I had to keep reminding myself that there is nothing inherently superior about colors derived from plant or animal sources using traditional methods. I tried to see colors through the eyes of the Turkish women who were working with them, as well as through the eyes of American rug buyers.

Ahmet Balci's wife, Rahime, is a weaver. She showed me rugs that she had made years ago for her dowry, all done with synthetic dyes, as that's all there was available then. The colors are neon-bright, especially the orange. On Rahime's floor was an older, synthetic-colored rug and a newer one done with plant-dyed yarns. She said that she still prefers the brighter rug, but that she was beginning to enjoy the softer colors, and when she weaves for sale, she uses *only* natural dyes. Business, after all, is business. ❖

Further reading:

W. Bruggemann and H. Bohmer. *Rugs of the Peasants and Nomads of Anatolia.* Munich, Germany: Kunst and Antiquitaten, 1983.

PHOTO BY AUTHOR

A weaver in a west Turkey village with a newly woven traditional rug.

DYEING IN CHINCHERO, PERU

BETTY DAVENPORT

WITH NILDA CALLANAUPA

My love for textiles and wandering along mountain trails has led me several times to Peru, where I can enjoy both at the same time. All throughout the Andes small villages are tucked into high mountain valleys. Here, the people still spin and dye yarn and weave all their traditional clothing. It was during a visit to Chinchero, a village on a high plateau near Cuzco, that I met Nilda Callanaupa.

Although Nilda has a degree in tourism from the University of Cuzco and has a job as a travel agent and trekking guide, she is at heart, first and foremost, a weaver. When she has spare time she takes her backstrap loom and climbs up the hill behind her apartment where she has peace, quiet and a good view over the red tile roofs of Cuzco as she weaves. Every weekend she is back in Chinchero with her parents and sisters, spinning, dyeing, weaving, experimenting and researching the older textile techniques of her people. She spends a lot of time talking with the older women in the village to learn how they used to knit the traditional caps and how they used to dye with plant material. Using plants to dye wool has been a tradition throughout the Andes for at least 3,000 years and was developed to a very high level of skill. However, since the introduction of aniline dye powder in the late 1800's, the weavers have found it much easier and more time-saving to simply buy the powder at the market to achieve the bright colors they like for their traditional weavings.

One natural dye that has remained very popular throughout the years in Chinchero is *nogal*, a kind of black walnut (*Juglans* sp.). Nilda explains that it is very difficult to achieve the light caramel color that is especially prized for men's ponchos. At least once a year every weaver in Chinchero will dye enough wool with nogal to make one or two ponchos.

Walnut trees do not grow at the 13,000-foot elevation where Chinchero is situated, so the weavers must buy the branches from families down in the Urubamba valley. The color in the leaves is strongest from June to September. When the supply of nogal arrives, the stems, leaves and hulls must be crushed and broken by pounding with a rock or pestle. A huge pot about two feet in diameter is filled with the crushed material and just enough water is added to partially cover them. A small fire is built under the pot and the nogal is cooked for several hours.

BETTY DAVENPORT *is a weaver, hiker and mountain climber who lives in Richland, Washington. She leads annual Weaver's Treks in the highlands of Peru.*

Nilda's mother, Guadalupe, supervises the dyeing process. Some of the branches are arranged over the top of the cooked mass or across the top of the pot, making a rack on which to lay the wool skeins to hold them above the liquid. The skeins are dipped quickly in the dye liquid and laid on the branches above the steaming caldron. The skeins must be dipped, steamed and turned frequently to obtain an even color. While some are being turned, others are covered with more branches out of the pot in order to catch the steam. If a darker shade is wanted, the skeins can later be left in the warm dye liquid overnight. The dyers take great care to not let the yarn boil in the dye liquid or it will become too brown and too dark. The whole family works together to orchestrate the process. No wonder this is a once-a-year event!

Although the people in the community generally know that plants other than nogal can be used to dye yarns, Nilda hadn't actually seen anyone doing it. When she was in her teens and learning the more advanced weaving designs, she became interested in learning to make dye from plants. She talked to the older women who still remembered how they used to dye yarns and experimented on her own, developing the following dye methods:

Nilda uses a small clay pot and builds a fire under it out of whatever dry branches and stems are available. The fire is small and needs to be stoked often. From experience, she knows just how much water and dye material is needed to dye one skein of yarn. Generally there is a high proportion of dye material to wool so the colors are strong, brilliant and fast. The skein usually consists of two hand spindles' worth of yarn. The yarns from the two spindles are wound together into a skein for dyeing. The plants she uses yield substantive dyes and do not need mordants to fix the color.

One day I was in Chinchero when Nilda's neighbor, Engracia Quispe, was getting ready to dye some yarn a brilliant red with cochineal. Peru is one of the world's sources of cochineal bugs. The bugs grow on the opuntia cactus in several regions north and west of the Department of Cuzco and can be purchased in the markets in Cuzco. Engracia measured out a handful of dried cochineal for her skein of yarn and ground them up with a pestle. She put the ground bugs in a cloth bag and boiled it for about one hour. She then squeezed the juice of about ten Peruvian limes into the pot. The acid of the lime juice shifts the color to bright red; without the lime juice she would get a magenta color. She put the wet wool skein in the dye and cooked it gently for 10-15 minutes. She left the pot on the fire until the fire died out and left the wool in the dye liquid for two days before rinsing it in clear water. The result was a bright carmine red. She could use the dye liquid again several times for lighter colors until it was completely exhausted.

Later as we walked around Chinchero and clambered up and down the Incan terraces on the way to visit the nearby ruins, we saw lots of bright yellow flowers blooming that Nilda calls *kiko*. It is a species of *Bidens* but is often referred to as coreopsis. Kiko blooms wild around Chinchero from March through June. Nilda explained that she only picks the flower heads and that the color for dye is strongest when the flowers are most plentiful in April and May. She generally uses an equal weight of flowers to wool in order to get a brilliant golden yellow. She adds enough water in the clay pot to cover the flower heads and cooks them for one hour. She leaves the flower heads in the dye when dyeing skeins, but removes them when dyeing fleece. She lets the wool simmer in the dye at least 15 minutes, then lets it cool in the liquid for several hours before rinsing in clear water.

Another day as we were having a picnic lunch along the roadside, Nilda pointed out a bushy shrub behind me that had small, shiny oval leaves and lots of one-half-inch-long thorns. This is *ch'ecchi*, a species of barberry (*Berberis* sp.). Nilda explained that the bright orange root gives a greenish-yellow dye. After lunch we dug up a hunk of the sweet-potatolike root from the rocky soil. Nilda said that it must be used fresh and is best the day it is dug. Back in Chinchero, Nilda mashed the root into small pieces with a pestle and boiled it in water for about 45 minutes. She strained the pieces out of the liquid before adding the skeins. Sometimes she bundles the mashed root in a cloth to contain it and leaves it in the dye pot. She cooked the wool gently for 10-15 minutes and left it in the dye liquid overnight before rinsing in clear water.

After dyeing, Nilda or her mother will spin the doubled yarns in the reverse direction to ply them, adding a great deal of twist that is needed for their durable warp-faced fabrics. After plying, the yarn is wound into a tight ball and is stored away for a while to allow the twist to set. The dyed yarns are used to weave the colorful designs typical of her village in *ch'umpis* belts, *llijllas* (shoulder cloths) and *k'eperinas* (carrying cloths). Nilda encourages other weavers in her village to carry on the tradition of spinning wool yarn and dyeing it with plant dyes instead of weaving with the brilliant, inexpensive acrylic yarns that are so readily available in the local markets. ❖

PHOTO BY AUTHOR

Engracia Quispe, Chinchero, Peru, dyes with cochineal.

Indigofera suffruticosa

ELIZA LUCAS PINCKNEY

CULTIVATOR OF INDIGO

MARCIA BONTA

Colonist Eliza Lucas Pinckney loved "the vegetable world extremely," and experimented with growing ginger, cotton, alfalfa and cassava, hoping to bring them "to perfection." But she is most famous for her experiments in indigo culture which she began when she was 17 years old. Through her persistent efforts indigo, which previously had had to be imported by the English from the French West Indies, became a valuable cash crop for the colony of South Carolina.

Born on the island of Antigua in 1722, Eliza was educated in England under the care of her parents' friends. Her father, George Lucas, was a sugar planter and lieutenant colonel in the British Army, stationed in Antigua, but when he inherited three South Carolina plantations from his father, he moved his semi-invalid wife and two daughters to a 600-acre plantation along Wapoo Creek, six miles by water from Charleston and 17 miles overland. From there he also directed the operations of 1,500-acre Garden Hill on the Combahee River, and the 3,000 acres of rice-growing lands he owned along the Waccamaw River. Lucas hoped that the change in climate would help his sickly wife. He also feared the

MARCIA BONTA *is a naturalist and writer who lives in Tyrone, Pennsylvania.*

Spanish and thought South Carolina might be a safer place to live.

The Lucases had barely begun their new life in South Carolina when, in 1739, hostilities with Spain erupted and Major Lucas was recalled to Antigua, leaving his eldest daughter Eliza in charge of all three plantations. Despite her youth, she was happy in her new, challenging life, following a rigorous daily schedule that appalled many of her more traditionally-minded female neighbors.

Every morning she arose at 5:00 a.m. and read classical writers such as Plutarch, Locke and her particular favorite Virgil whose "calm and pleasing diction about pastoral and gardening" seemed comparable to what she experienced during a South Carolina spring. Two hours later, she walked in the garden, supervising the servants at their work and setting out the tasks for the day. Then she had breakfast followed by an hour of practicing the harpsichord. The rest of the morning she devoted to French, shorthand and teaching her younger sister Polly, as well as two little servant girls, how to read. Dinner was followed by another hour of music and several of needlework and her evenings were occupied by more reading and letter writing.

But her most absorbing work was horticultural — planting a fig orchard, a cedar grove and a "large plantation of oaks," as well as concentrating on making indigo a viable export crop for South Carolinian planters, a project others had tried before her but had not succeeded in. Every year she reported her progress in growing indigo to her father who supplied her with seed. By June of 1741, she was confident of eventual success if he would send seed in time for her to plant it by the latter part of March so the crop would mature before frost.

Her father complied and also sent her Nicholas Cromwell of Montserrat, who was an expert "indigo maker," as the head man was called, but both he and then his brother purposely spoiled the dye in 1742 and 1743 during the crucial fermentation period — probably because they did not want South Carolina to rival their native island in indigo production. Finally, in 1744, Lucas sent out an unidentified black man from the French Indies to serve as "indigo maker," and the Wapoo plantation produced its first successful dye cakes that year.

To produce indigo, the soil had to be carefully prepared. Once the bushes started growing they had to be watched until the crucial moment, just before blooming, when the leaves were ready to be cut. They were then steeped in large vats of water open to the sun until they fermented and turned the water a greenish color. This fermentation process took several days and was observed around the clock by teams of men directed by the "indigo maker" who never left the premises. When he decided the process was complete, the solution was strained and poured into a second vat, along with a small amount of limewater, where it was beaten with paddles until it began to thicken. This agitation determined the final color of the dye, since the longer the mixture was beaten, the darker the color became. Finally, the solution was put into a third vat and allowed to settle before the clear water was drawn off, leaving a sediment that was formed into dye cakes. After the cakes were carefully dried in the shade, they were ready for market.

While Eliza struggled with indigo production, she continued to send progress reports to her father on other experimental crops — cotton, "Guiney" corn and ginger — all of which had also been "cutt off by a frost." In addition, she wrote, the "Lucern [alfalfa] is yet but dwindlering."

Although she liked her solitude, she did accept as many invitations from the Charleston gentry as she could find time for. She was particularly fond of Elizabeth

Pinckney and her niece, who was visiting from England, and of Elizabeth's husband, Charles, who supplied her with books.

Suddenly, in January of 1744, Elizabeth, who had been an invalid for years, died of an enlarged spleen. At the same time, George Lucas, who had been made lieutenant governor of Antigua, sent his son George to Wapoo to bring his wife and daughters back to the West Indies. But, to everyone's surprise, Charles Pinckney quickly proposed to Eliza, and she as quickly accepted. By May of 1744 they were husband and wife, and Eliza entered the most joyful years of her life. Despite the age disparity (he was 45 and she was 22), it was a love match. Charles was a successful lawyer and planter who found time to talk to Eliza about her reading and to join her in her horticultural interests. He also possessed, according to Eliza, "a charming temper and disposition, gay and courteous manners, [was] well-looking, well-educated and of high religious principles."

She moved to Bellmont, the Pinckney plantation, which was only 15 miles from Charleston and overlooked a tributary of the Cooper River, but she, joined by Charles, continued to oversee the cultivation of indigo on both their own and her father's plantations. That summer they gave indigo seed to any planter who promised to raise and export it, and Charles sent six pounds of the blue dye made from the Wapoo crop to South Carolina's London agent, James Crokott, for an expert opinion on its quality. It turned out to be excellent, a worthy substitute for the French West Indies product.

The following year indigo valued at 225 English pounds was exported to London from the Lucas Garden Hill plantation alone. Other planters who had obtained seed from the Pinckneys were equally successful. After that, production of indigo by South Carolina planters skyrocketed. In 1746 135,000 pounds of the blue dye cakes were shipped to England. Two years later British merchants petitioned the parliament to put a six pence bounty per pound on indigo grown in South Carolina to encourage direct shipping to England. By 1754 217,000 pounds were being exported and careful indigo planters doubled their money every three years. Eliza Lucas Pinckney had fulfilled her primary horticultural goal: to make indigo a profitable crop for South Carolina.

She went on to live a full, long life, most of it in her beloved South Carolina, managing the estates left to her after her husband died of malaria in 1758, always experimenting with new crops and planting new trees. Her two sons also inherited her love of experimenting with horticulture. When the family was caught up in the American Revolution, they never wavered in their support of the patriots, despite strong ties to English friends and to American royalists, and both sons fought in the war on the side of the colonists. Later, Charles Cotesworth Pinckney, who became a prominent lawyer, was sent as one of South Carolina's delegates to the Constitutional Convention in 1787, and Thomas Pinckney became governor of South Carolina. He was also the man who negotiated the Pinckney Treaty with Spain.

No doubt it was the prominence of her sons, and not her own early accomplishments in indigo growing, that prompted George Washington to ask to be a pallbearer at her funeral on May 26, 1793. But historians have not forgotten the role of the young Eliza in the economy of South Carolina. As Edward McCrady later wrote in his *History of South Carolina Under the Royal Government,* "Indigo proved more really beneficial to Carolina than the mines of Mexico or Peru were to Spain...The source of this vast wealth...was the result of an experiment by a mere girl."

PHOTO BY AUTHOR

An indigo solution is being beaten to oxidize it and turn it blue.

A Blue Future for Mexican Indigo

Gary N. Ross

Having been born in Louisiana and educated as a field biologist there, I grew up with an indigo heritage. Louisiana once relied on the production and exportation of the dye for its economic survival. As a boy, I saw indigo plants thriving in cattle pastures throughout the state. These plants, I learned, had escaped cultivation more than 100 years earlier and had continually reseeded themselves. My interest in indigo was rekindled years later when, as a collector of textiles in the highlands of southern Mexico, I learned that several Zapotec weavers were still carrying on a tradition of natural indigo dyeing with their tapestry art.

In pursuit of the dye, in the summer of 1982, I consulted several weavers in the Zapotec village of Teotitlan del Valle who used natural indigo to produce a multitude of bluish, greenish and bluish-red hues in their elaborate handmade tapestries and rugs. These weavers did not prepare the dyestuff, called anil, but purchased it at the market town of Tehuantepec in the form of blue chalky cakes.

And so, to further my investigations, I left the dry, mile-high Oaxacan plateau and drove the tortuous mountain road down to the coast and the bustling town of Tehuantepec. There I learned that anil was still grown, but in only one village, Niltepec, which according to Zapotec legend means "place of the anil plants." Furthermore, I learned that for the last two years only the Tolentino Meza H. family had continued the tradition, selling the entire production from their home to the few weavers who came to purchase it.

Continuing my quest, I easily tracked down Niltepec, two hours southeast of Tehuantepec, and after a few inquiries, I located the Meza family residence. The senior Meza, a short, weather-beaten man of 61, called *maestro* by all, was the expert on indigo production. But Meza's news of his crop that year was not good.

The tiny seeds had been hand-sown in late April at the beginning of the summer rainy season, and they had sprouted on schedule. But the life-giving rains had failed to come, and the plants were small and weak. As I prepared to return to the Oaxacan plateau, we agreed that Meza would contact me if and when he harvested the crop.

Responding to a telegram from Meza, I returned to Niltepec in October. My reunion with the Meza household was affable, but filled with apprehension. The indigo plants were stunted. We headed directly to the fields, where I saw stiff-stemmed shrubs about four to six feet tall, with dull green feathery leaves, and with clusters of small coppery-pink flowers resembling those of pea and bean plants. I immediately identified them as *Indigofera suffruticosa*, the New World indigo plant, the same as that in my home state.

We sat on the baked earth while Meza explained the sequence of activities that I would soon observe, and that had taken place for as long as he could remember. The onset of flowering, he said, signals the time to harvest — a routine that usually continues on a daily basis for approximately four weeks but that this year would be much shorter. The plants are cut with a pruning knife a few inches above the ground during the hours just before dawn when the sap content in the plants is at its greatest.

The plants are bundled together by the dozen with strips of vine, loaded onto an oxcart and transported to nearby tanks: a set of two open vatlike structures with thick brick-and-cement walls, approximately nine feet square and from four to seven feet deep. Both are built where they will receive the full light of the sun, positioned side by side with a common wall, and with one tank on a sizable mound to allow it to drain easily into the tank below. An adjacent well provides adequate fresh water, which is hauled up by hand and poured into the upper vat.

The bundled indigo plants and water are added to the upper tank until it is one-third to one-half filled, taking about 96 bundles of plants — the complete harvest of one day. Several long wooden poles are jammed across the top of the tank to keep the plants submerged at least six inches below the water's surface, and they are left to steep in the hot tropical sun. During the day, the water warms and the plants' sap bleeds into the water, gradually turning it a light yellow-green. As time passes, air bubbles form among the submerged plants and percolate to the surface — the sign of fermentation. By night, the fluid is so effervescent that the tank sounds like a giant glass of Alka-Seltzer. Then, near midnight, the solution fizzles out, and a watchman, who has been periodically checking the progress, pulls a plug from the top tank to let the now bluish-green fluid drain into the lower tank. He then goes to the village to awaken two comrades, who return with him for the next stage of the process.

By now the tank has drained. (The spent plants are later given to the town's rummaging pigs as fodder.) The two men take up positions along one side of the lower tank. Each holds, in javelin fashion, a long pole with a shorter crosspiece that forms a T. These poles, or beaters, are forcefully thrust and pulled in unison within the liquid, setting it into violent motion. After five or six hours, sufficient air has been introduced to transform the color of the murky fluid to the intense blue associated with the name indigo. (In chemical terminology, the indigo has become oxidized.) In the process, the human beaters are splashed with the bright stain of blue, marking them with their trade.

The dyestuff next must be separated from the water by sedimentation. This is achieved by adding a concoction of water and the pulverized berries of the *gulabere*, a medium-sized, mulberrylike tree that is common in the area. The tree produces small clusters of small, greenish fruit in August that contain a clear slime. When added to the indigo solution, the mucilage acts as a catalyst and causes tiny flecks of solid blue — indigo dye — to form. Within 15 to 20 minutes, the solution has separated into a clear watery upper layer and a lower one of blue sediment — probably the most beautiful ooze in nature. After the fluid is carefully drained through the tank's escape vent, the one- to two-inch layer of blue sediment is swept from the tank through the same vent, collected in ceramic jugs, and carried off to a nearby shed.

This airy, open-sided, grass-roofed structure is equipped with two wooden racks holding squares of cotton cloth. The blue sediment is slowly poured through a handwoven cane sieve to filter out twigs and pieces of leaves, and into the suspended cloths, which permit the water to drain slowly. By late afternoon the dripping has slowed substantially, and the bags are bound and transported on poles to the security of the owner's open-air patio. Early the next morning the dye, which now has the thick smoothness of freshly prepared pudding, is spread with a wooden spatula onto clay roofing tiles and placed in the sun to dry for eight days. The months of cultivation, the weeks of back-breaking harvest and processing, all are reduced to hard cakes of chalky blue — pure indigo dye.

At least, that was the way the system had worked in the past. The results in this year of drought were still to be seen. The following day, at three o'clock in the morning, we assembled in the first of three small cultivated fields for a projected two weeks of processing. But the cut plants appeared much less succulent than usual, and that afternoon the steeping process proceeded very slowly, not ending until five o'clock the following morning instead of the customary midnight. Also, only three hours of subsequent beating were required to oxidize the solution instead of the usual six or seven.

To further complicate matters, the *gulabere* trees were now, in mid-October, long past their prime, and berries were scarce. An alternative solution from the leaves of a common low-growing weed called *malva* (a kind of mallow) was prepared, but for some unknown reason, it did not work, and the usually clear water that drained from the lower tank into the nearby stream was, instead, an intense blue, leaving behind only a very thin sheet of sediment that at day's end was contained in two cloth bags. The second day duplicated the first. Time, labor, and potential capital all literally were washing down the drain.

For the next 12 days the procedure was repeated, the outcome always the same. In the end, a paltry 25 pounds of dried dye was the result instead of a typical average of ten times that amount. The value of their month's labor: $500. We concluded that the plants simply were too dry from the lack of rain during the growing season.

Given this outcome, Meza and his family decided not to plant indigo again and risk further calamities. They would switch entirely to cattle ranching, a proven, stable and profitable occupation for the area. In the spring of 1983, however, the unexpected occurred. Isaac Vasquez, one of the Zapotec master weavers from Teotitlan del Valle, visited the Meza family to convey a plea from his fellow artisans: "Continue to produce indigo and we will pay whatever is necessary; no other source exists except from the foreign market."

New seeds were sown immediately. But once again the rains were inadequate, and the harvest was meager — only slightly greater than in 1982. In 1984 Vasquez petitioned again: once again the fields were planted, and once again the rains failed. By 1985 the total production had hit rock bottom — 11 pounds of dried dye, for which the artisans paid $600.

And so it goes. Mexican indigo, a unique Zapotec tradition, now teeters on the brink of extinction, remaining alive each year only because a handful of individuals are too proud of their past to concede to what must surely seem the inevitable. ❖

A LONGER VERSION OF THIS ARTICLE FIRST APPEARED IN ***AMERICAS***, PUBLISHED BY THE GENERAL SECRETARIAT OF THE ORGANIZATION OF AMERICAN STATES IN ENGLISH AND SPANISH. REPRINTED WITH PERMISSION.

MAYO INDIGO

Susan Skirvin

For thousands of years, rich blue colors of indigo have been prized by people around the world. The chemical compound which produces indigo blue occurs in several plants of different botanical families. Long ago, dyers living where indigo-bearing plants grow developed methods to extract and use the dye. Inside the plant, indigo pigment is colorless and concentrated in the leaves: A torn leaf turns blue at the edge where pigment reacts with air. This blue will stain but not dye yarn permanently: It doesn't dissolve in water so it stays on the yarn's surface. If the blue indigo pigment is dissolved in an alkaline, chemically reduced solution (one with extra oxygen removed), it makes a yellowish-green dyebath which will dye many types of fibers. The dyer immerses clean, wet yarn in the indigo dyebath and lets it soak thoroughly. When the yarn is first pulled out, there is only a pale yellowish-green color, but as air reaches the fibers and reacts with the dissolved indigo, the yarn turns blue. Repeated dips into the dyebath, each followed by airing, produce deeper blues which are very resistant to fading by light and washing.

I have grown indigo (*Indigofera suffruticosa*) in my backyard. This plant is a tropical bush, in the legume family. I obtained seeds from a nonprofit seed-conservation organization called Native Seeds/SEARCH, which is dedicated to preserving genetic diversity of crops of native people of the arid Southwest. These tiny black indigo seeds come originally from the Mayo people of southern Sonora and northern Sinaloa in northwestern Mexico, where they live along lowland river valleys and in dry desert areas between valleys. Mayo craftswomen use *Indigofera suffruticosa* and other native plants to dye handspun wool yarns which they weave into thick blankets and rugs. No one knows for certain whether prehistoric people in this area used indigo before the Spanish invasion, but it has been used ever since. Fine old historic Mayo textiles contain both indigo- and aniline-dyed yarns; the aniline dyes have faded badly, while the indigo is still beautiful blue.

The history of Mayo weaving is not as well known as that of the Navajo people in the United States, but probably it followed a similar course: Early settlers and traders bought from the Mayo and encouraged their weavers to keep spinning, dyeing and producing textiles for sale. The completion of the first railroad through Mayo country about 1918 brought many tourists who eagerly bought the colorful blankets. However, by World War II weaving was in a decline which continued to the early 1970's, when only one in twenty blankets contained any vegetable colors.

A revival of plant-dyed blankets was en-

Susan Skirvin *has been using natural dyes for 15 years, teaches and is a graduate student in geoscience.*

couraged by interested Americans, including Barney T. Burns of Native Seeds/SEARCH, who buys Mayo blankets to resell in the United States. Mayo weavers said that indigo had become hard to find, since it had grown on sandbars along rivers which had since been dammed. Fortunately, a large indigo bush in the village of La Bocana provided seeds which were collected by Native Seeds/SEARCH and distributed to weavers for them to grow.

As you might expect of a plant coming from tropical riversides, *I. suffruticosa* loves heat and water. To grow it in the hot, dry Arizona desert, I plant presoaked seeds in the garden after frost is clearly over, usually in late March or very early April. Young plants need time to develop good root systems before June; then they love the heat, as long as they get plenty of water. By August the plants are nearly four feet tall and starting to bloom, ready to harvest and dye. In Mayo country, frosts are rare and dyers can harvest year-round; in my yard indigo slows down around October and dies with the first killing frost.

The Mayo use a "direct-dye" method to extract pigment and prepare a reduced, alkaline dyebath in one step: Freshly harvested plants are fermented in urine for several days. Fermentation bacteria create reducing conditions, and urea in urine adds necessary alkalinity to dissolve indigo from the soaking plants. Yes, it does smell, but indigo itself has a characteristic strong odor anyway, and this method is simple and straightforward. I tried a variation of direct-dyeing with my indigo plants, using tap water and letting summertime heat do the fermenting. Since my water is alkaline already, I thought it would dissolve the reduced indigo. It did, a little, but the dyebath was blue-green instead of yellow-green and much of the color washed off my yarn. That dyebath was not alkaline enough; I could have added ammonia or

PHOTO: RITA BUCHANAN

The yellow-green color changes to blue when the yarn is exposed to air.

wood ashes to dissolve all the indigo. I still love the colors, though: light sky-blue to medium indigo-blue with a touch of grey on wool, cotton and linen, and a subtle lavender on silk, from three dips and airings. Maybe I'll try the Mayo method next summer, for stronger, clearer blues from the garden. ❖

Special thanks to Barney T. Burns and Mahina Drees of Native Seeds/SEARCH for information on their work with Mayo weavers.

FOR FURTHER READING:

Fontana, Bernard L., Edmond J. B. Faubert, and Barney T. Burns. *The Other Southwest: Indian Arts and Crafts of Northwestern Mexico*. Phoenix: The Heard Museum, 1977.

MUSHROOMS FOR COLOR

MIRIAM C. RICE

ILLUSTRATIONS BY DOROTHY BEEBEE

Dermocybe phoeniceus var. *occidentalis*

Mushrooms are a reliable organic source of beautiful dye colors — reds, oranges orange-reds, yellows, greens, blues, purples, blue and red-purples, blacks, and tans, grays and browns galore. The subtlety and variety of possible shades is endlessly fascinating, yet a full rainbow can be attained from as few as three or four species. Mushrooms provide beautiful colors which are easy to extract and use, and the colors are washfast and lightfast. Mushroom dyes can be used on protein fibers such as wool, silk, camel hair, llama, alpaca and furs; and on cellulose fibers such as cotton, rayon, raffia and reed to give colors of varying intensity and effect.

I plunged into exploring and experimenting with mushrooms for dyeing nearly 20 years ago. The excitement of discovery and frenetic recording of hundreds of mushrooms and their colors culminated in the publication of *Let's Try Mushrooms for Color* in 1974. This has sparked interest in more and more countries. (About 70% of mushroom species occur worldwide, so mushroom dyers in different places can easily share their experiences and observations.) *Mushrooms for Color*, the expanded 1980 edition of my book, is the basis for all books on the topic in other languages, including Swedish, Danish, Norwegian, German

MIRIAM C. RICE *is the author of the book* ***Mushrooms for Color***.

and Finnish, and an article in Japanese. Since 1980, five international symposia have provided an opportunity for textile artists, mycologists and mushroom cultivators to exchange ideas and exhibit their work.

A mushroom is the spore-bearing or reproductive part of a fungus organism. Underground, the same organism is made up of a network of threadlike hyphae which grow, spread, digest and store up nutrients derived from the dead leaves, duff, pine needles, old wood and other organic debris in the soil. Now more than ever, fungi are appreciated and recognized for the powerful function they perform in sustaining life as we know it and maintaining the health of the planet as a whole. In particular, mushrooms play a remarkable role in forests. Fungal hyphae interact with tree rootlets to form partnerships called mycorrhizae; these mycorrhizae can be traced throughout the forest floor as scientific proof of the interconnectedness of the whole forest ecosystem. By inoculating seedlings with proper fungal spores in the nursery, small trees when planted have with them the partner that helps obtain minerals from the soil. For example, *Pisolithus tinctorius* has been used for years in reforesting strip-mined areas with lodgepole pine. The same fungus bears fruiting bodies which are an excellent dye source, yielding rich browns to blacks.

Sarcodon imbricatum

Just as mycorrhizal fungi are now cultivated in nurseries, mycologist/cultivators such as Paul and Cruz Stamets, who operate "Fungi Perfecti" in Olympia, Washington, have developed techniques for cultivating more and more species of edible mushrooms. This research may soon extend to growing the dependable species of dye mushrooms.

Not all mushrooms give color. The common white mushrooms from the grocery store give only tans, beiges and off-whites. For better colors, try wild species gathered from the woods and fields. Learn to identify what you find by studying reference books (see list at end of article), joining a mycological society or taking a mushroom identification class at a community college, botanic garden or nature center. Learn to recognize the poisonous species and avoid them if you prefer; in any case, remember always to keep your dyeing equipment and materials separate from your cooking and eating activities.

The process of dyeing with mushrooms is the same as with other natural dyestuffs such as leaves, roots, bark, flowers and berries. To make a dyebath, simply cover the mushrooms with water, heat to simmering temperature and simmer one hour. You can strain off the liquid before proceeding to dye, or not, as you wish. As a general rule for dyeing wool, start with one pound of natural dyestuff per pound of fiber. However, the amount of pigment in different species of mushrooms varies, so you may need to use more or less to get the color desired.

Premordanting the fiber assures clearer color and more even results. To premordant wool, dissolve the appropriate amount of mordant in water and add to pot with enough water to cover wool, slowly heat to simmering temperature

DERMOCYBE (CORTINARIUS) PHOENICEUS VAR. OCCIDENTALIS

WHEN FOUND: fall to early spring
WHERE FOUND: mixed conifer forest

MORDANT	DRIED	FRESH
None	Apricot	Apricot
Alum	Rose	Wine Red
Chrome	Burgundy	Burgundy
Tin	Red	Blood Red
Copper	Grayed Purple	Grayed Burgundy

SARCODON (HYDNUM) IMBRICATUM

WHEN FOUND: fall
WHERE FOUND: in moist conifer forests or mixed conifer hardwoods (huckleberry-manzanita)

MORDANT	OVER RIPE/FRESH	DRIED
None	Warm Gray	Gray
Alum	Green-Blue	Gray
Chrome	Gray-Green	Blue-Gray
Tin	Blue	Blue
Copper	Forest Green	Green-Gray

Note: Blue dye only if mushroom is very mature-old, even slimy. No blue from fresh, young mushrooms.

PHAEOLUS SCHWEINITZII

WHEN FOUND: fall and winter
WHERE FOUND: in clusters as a parasite on the ground, from roots of dead stumps or on live or dying conifers

MORDANT	FRESH NEW GROWTH	FRESH OLD GROWTH
None	Yellow	Burnt Orange
Alum	Yellow-Orange	Burnt Orange
Chrome	Burnt Orange	Red-Brown
Tin	Orange	Red-Brown
Copper	Red-Brown	Red-Brown

and simmer one hour. Remove the wool, rinse it well and use immediately or dry for later use. For color samples, I prepare swatches of six strands of the premordanted yarns, knotting each in such a way as to know at a glance or feel which mordant was used. (I use 0 knots for no mordant, 1 for alum, 2 for chrome, 3 for tin, 4 for copper and 5 for iron.) Having a ready supply of these swatches on hand makes it easy to find out how each new fungal find might reward you when lifted from the simmering brew.

Add the premordanted yarn to the dyebath and simmer one hour. Remove the fiber after the dyebath has cooled, or leave it in the dyebath overnight — this seems to intensify the color and improve the lightfastness. Then rinse the fiber, wash it with soap or detergent, rinse again and hang to dry. Voila! The dyed yarn will not run or bleed.

You can re-use the dyebath by adding more fiber. Mushrooms often contain more than one pigment, so a different color sometimes results from a second dyeing in the same dye bath. The first dyeing depletes one color, leaving the less dominant color to appear. For example, *Sarcodon imbricatum* gives blue in the first dyeing, and brown in subsequent dyeings. Also, the palette supplied by water-soluble fungus pigments is immensely enlarged by using different mordants, mixing different mushrooms, overdyeing and adjusting the pH of the bath. Adding vinegar to reduce the pH of the dyebath, or ammonia to raise it, can produce dramatic color changes. Read books and experiment to learn more about the possibilities of using mushrooms for color.

When all the dye has been extracted, what is there to do with the leftover depleted mushrooms? I use them to make paper. The variety of textures, color and translucency is fabulous. There is mushroom-dyed paper, and there is paper made from the mushrooms themselves...beautiful, different and with untold possibilities as a source for paper other than our trees. ❖

Detail of "Fragile Flight" tapestry. *Opposite page:* Detail from "Forest Floor," another mushroom-dyed tapestry.

PHOTO: T. ELLEN ATHENS

FOR MORE INFORMATION:

Arora, David. *Mushrooms Demystified.* Berkeley, CA: Ten Speed Press, 1986.

McKnight, Kent H. and Vera B. *Mushrooms: A Peterson Field Guide.* Boston: Houghton Mifflin, 1987.

Miller, Orson K. Jr. *Mushrooms of North America.* New York: E.P. Dutton, 1979.

Rice, Miriam. *Mushrooms for Color,* illus. Dorothy Beebee. Eureka, CA: Mad River Press, 1980.

ARCTIC LICHEN DYES

Wendy Chambers

Lichens are some of the most important colonizing plants, yet some of the least studied. An amateur botanist and dyer can still make exciting discoveries about their distribution patterns or chemical properties. So if the subtle and the inconspicuous draw your attention as they do mine, lichens will offer you a world of wonder. Their varied appearance gives no indication of the riveting colors which they yield to the dedicated dyer.

The first step in dyeing with lichens is learning to identify them by species. (Each species of lichen actually combines two organisms, an alga and a fungus, in a symbiotic union.) Mason Hale's book *How to Know the Lichens* (Wm. C. Brown, Dubuque, Iowa, 1979) is an excellent reference to consult. Basically, lichens are classified into three growth forms: crustose, foliose and fruticose. Crustose lichens grow in a thin layer, foliose lichens have a leaf-like body attached to the substrate by a "rhizome", and fruticose lichens may be hairlike, shrubby, fingerlike or strap shaped. Lichens are further characterized by various structures which appear on their surfaces. You'll need to use a magnifying glass to distinguish the different kinds of protuberances, pits and other features and to correctly identify a lichen.

Lichens are slow-growing, particularly in the high latitudes where long periods of dayless nights, cold temperatures and lack of moisture limit growth to a few days a year. It has taken thousands of years for the arctic to be colonized by these plants which are now stressed by atmospheric pollutants brought on air currents from industrialized Europe. It's very important for lichen collectors to think about conservation. Don't collect in the same spot more than once in five years; in some areas the interval should be not more than once in a human lifetime. To allow for regeneration, remove less than one-third of the growth. Identify and experiment with a small sample of each new species before collecting in quantity. Try to collect a species only in its most favorable habit where growth is abundant, not where it is rare.

Wendy Chambers *lives in Whitehorse, Yukon, where she is proprietor of Down North, a fiber business specializing in musk ox qiviuq. She teaches workshops in spinning, weaving, and knitting qiviuq; and in lichen dyeing.*

By understanding the chemistry of lichens and applying the principles of color theory, you can obtain a satisfying range of colors from the common lichens in a collection area. Each region offers a distinctive palette which can be obtained without endangering rare species.

Dyeing with Lichens

Colors are obtained from the acids contained in the center layer of the thallus, or body, of lichens. Different acids produce characteristic colors. After I have identified a lichen I consult a reference book such as John Thomson's American Arctic Lichens: *The Macrolichens* (Columbia University Press, New York, 1984) to determine which acids are present and to predict which colors the lichen may yield. As some lichens contain more than one acid, they may yield a range of colors with different methods or treatments.

One thing to note is that lichens produce substantive dyes. That is, the color molecules from the lichens can bind to the molecules of protein fibers such as wool or silk without the aid of a metallic salt. This eliminates the mordanting process; in fact, lichens themselves can be used as a mordant with other vegetable dyes.

There are two basic processes for extracting colors from lichens. The first uses boiling water. Clean the lichens of all dirt and foreign vegetable matter, put them in a pot with some yarn and add warm water. Heat the dyebath to a simmer and hold until the desired shade is achieved. Combining the lichen and the fiber in one pot is called contact dyeing. It is messy but yields stronger, clearer colors than cooking the lichen separately, straining off the liquid, and then adding the fiber for further simmering.

Fermentation is the second method of obtaining colors from lichens. It requires steeping the lichens in ammonia, either full strength or diluted with water, for a period of several days or weeks. This produces the dyes traditionally called orchils, which can impart pinks, fuchsias, and lilac colors to the fibers by processes of cold steeping or hot simmering. With some lichen acids, this process is taken a step further. The yarn is cold soaked for several days until it turns a deep pink, and then, while still damp, exposed to strong sunlight which oxidizes the pink color to blue.

Two supplementary steps can be used to extend the range of colors. First is overdyeing, or dyeing the yarn twice in different dyebaths. Second is shifting the color by adding small quantities of chemical color modifiers to the dyebath, usually at the end of the dyeing process. Copper sulphate will shift yellows to greens, and adding more ammonia will intensify the green or make it mossier. Iron will darken colors and push some tans to darker brown or gray. Acetic acid (vinegar) will shift some orchil pinks to more reddish hues, while washing soda (sodium carbonate) will shift them to more lilac or bluish hues.

One Dyer's Palette

I depend on just a few lichens to produce the primary colors of red, yellow and blue. *Cetraria delisei*, which is very common on the southwestern coast of Banks Island where it grows in large cushions, yields beige to dark brown by the boiling-water method. Using the fermentation method on the same lichen releases an orchil acid which produces a strong fuchsia color that can be shifted to a more reddish tone with acetic acid. This lichen doesn't react strongly to the addition of soda, but *Dactylina arcticum*, another orchil-producing lichen which is abundant in some areas, does react to soda.

Thamnolia subuliformis and *T. vermicu-*

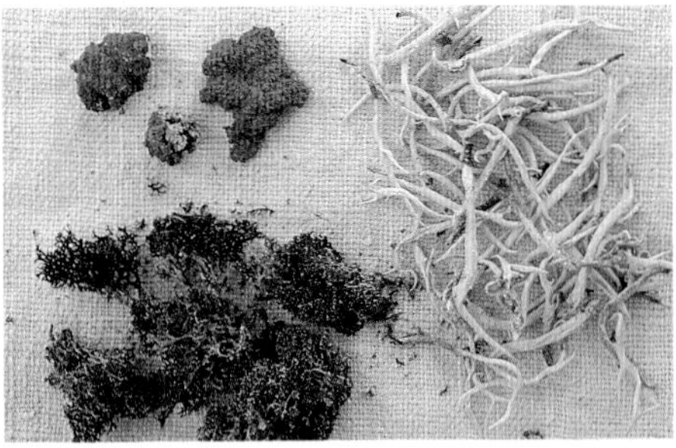

Lichens, clockwise from top left: *Xanthoria, Cetraria, Thamnolia.*
PHOTO BY RITA BUCHANAN

laris may appear scattered like thin white worms amongst the other tundra lichens, or may grow in dense mats where each strand is thicker and more branched so that a handful collected looks like a stack of bleached caribou antlers. Using the boiling water method on thamnolias gives colors varying from clear lemon yellow to a warmer, apricot tone. *Cetraria telesii* will dependably produce fluorescent yellows. To obtain blue, I ferment and oxidize *Xanthoria*, a crustose lichen easily recognized by its bright orange color and preference for rocky bird perches.

For the secondary colors — green, orange and purple, I use overdyeing. The photochemical oxidation of *Xanthoria* blue on yarn previously dyed yellow with *Cetraria tilesii* produces a blue-green, while an afterbath of copper sulphate on the same yellow produces a fresh spring green which can be further heightened by adding some ammonia. The same color modifiers on yellow dyed from *Thamnolia* produce mossier greens.

For orange, I start with various yellows from *Thamnolia* or *Cetraria tilesii*, then overdye with red from fermented *Cetraria delisei*, gradually increasing the strength of the dyebath until I achieve the desired shade. These oranges are very rich apricot and salmon shades.

Purples are obtained from the photochemical oxidation of *Xanthoria* blue over yarn previously dyed fuchsia with fermented *Cetraria delisei* or *Dactylina arcticum*. An afterbath of soda on the *Dactylina arcticum* produces a lilac shade.

The soft subtle tertiary shades can be achieved by starting with a yarn dyed beige with *Cetraria delisei*. The beige acts as a toner on the overdyed colors. For instance, to make a moss green which is deeper and richer than the one described above, start with beige, overdye with yellow which has been simmered a long time to burnish the color, and then modify it with copper sulphate and ammonia. A lovely old rose can be had by overdyeing *Cetraria delisei* beige with fuchsia from a fermented bath of the same lichen. Smoky gray blues are the result of oxidizing *Xanthoria* on beige, brown or light green yarns.

A lichen collector becomes deeply immersed in the landscape, as it requires keen observation to spot a specific lichen and a thorough understanding of topography to locate the right habitat for each species. This attention to detail extends to the dyeing process. You must understand each lichen's chemistry to coax it to release the colors hidden within. The dyeing process becomes an intimate relationship with the earth which, even in the lilliputian world of lichens, reveals many mysteries to the patient observer. ❖

PHOTO BY MARJORIE HUDSON

Yarns dyed with lichens.

BLUE DYES FROM *XANTHORIA* LICHENS

JUNE UPTON

The xanthorias are foliose lichens of bright orange-yellow hue, which grow on rocks by the sea and on the roofs of farm buildings. The species I use is *X. ectanoides*, which grows on granite rocks on the coast of Cornwall, England. In autumn, older patches of lichen become loose and detached, to be blown away in the first gale — if I don't get there first! I only gather small quantities of loose material, so as not to interfere with the growth pattern. A salty atmosphere and a sunny location produces the best dye material, and while other species of *Xanthoria* also give blue, none are of any use if collected from roofs made of asbestos, perhaps due to its mineral nature.

JUNE UPTON *is a handspinner and knitter who lives in Helston, Cornwall, England. She has been using natural dyes for nearly 20 years, employing plants and lichens which are plentiful in her area.*

PHOTOGRAPH BY AUTHOR

Lichen *Xanthoria ectanoides*, United Kingdom.

I follow the recipe given below. Dyers who are familiar with obtaining blue from indigo and woad will note the similarities of gentle heating and the occurrence of a color change (although the dye pigments involved are different compounds). The practice of obtaining a blue dye from this lichen seems to be of recent origin. Dyers who have done historical research into old dyes in the country can find no mention of this technique earlier than Eileen Bolton's *Lichens for Vegetable Dyeing*, published in 1963.

I use this lichen to dye wool fleece which I subsequently spin, but it can be used to dye skeins of wool yarn. The dye also works, but gives softer colors, on cotton and silk. It isn't necessary to pre-mordant the fiber, but using an alum mordant on wool may improve the lightfastness of the dye. This blue dye is best used for articles that are not exposed to much sunlight.

1. Crumble 2 oz. of dried *Xanthoria* lichen into a large glass screw-top jar. Pour in 20 oz. of a solution of 1 part nonsudsy household ammonia to 2 parts water. Cover the jar, and shake it several times daily.

2. After six weeks or longer, transfer the contents of the jar to an enamel or stainless steel pan. Add 1 tbsp. noniodized salt dissolved in 2 cups of boiling water. Bring to a boil, and simmer for 45 minutes with the lid off the pan. This is best done outdoors, as the boiling drives off the ammonia and causes strong fumes. Allow to cool, and strain off the liquid.

3. This step must be done on a sunny day. Put the dye liquid into a container large enough to hold 4 oz. of clean fleece, and add sufficient water to fully cover the wool. (Remember that in natural dyeing, success depends on the ability of the fiber to take up the dye, not on the dilution of the dyebath.) Stand the container in a bucket filled with boiling water. After 30 minutes, remove the wool to check its color — it should be a strong, "shocking" pink! Squeeze out the fleece and spread in out in a clear plastic bag. Don't allow it to dry yet.

Put the bag in a warm place exposed to sunlight. I find the best place is indoors on a sunny windowsill, and use a shiny metal tray under the bag as a reflector. When one side of the wool is "done," turn the bag over. The fleece may be dipped in the dye and blued in the sun several times for a deeper shade. When the color is deep enough, or the dyepot is exhausted, allow the wool to dry out of sunlight. When it is completely dry, wash in cool soapy water and rinse, adding 1 tbsp. of salt to the final rinse.

NOTE: Lichen pieces left over from the fermentation may be boiled with fresh water to give a dusky pink on chrome-mordanted wool. My family calls *Xanthoria* "Auntie's Magic Lichen," a name that originated when I was teaching my niece about plant dyeing. We discovered that this lichen reacts instantly to pH changes — going alternately pink (with alkali) and yellow (with acid). ✣

DYEING 1750's BRITISH REDCOAT UNIFORMS

JIM LILES

During the French and Indian War (1754-1763) the British Colony of South Carolina felt threatened by French activities in the Mississippi Valley. To counter this threat the Colony formed and sent The Independent Company of South Carolina (approximately 100 soldiers) to build and garrison Fort Loudoun, in the heart of the Overhill Cherokee Indian nation. The purpose was to reinforce the allegiance of the Cherokee with the British and to maintain trade between the Cherokees and the Colony of South Carolina. Thus, Fort Loudoun became the westernmost British outpost in the Southern American Colonies. The recreated fort, on its original location, lies on the Little Tennessee River (now Tellico Lake), about 40 miles southeast of Knoxville, Tennessee.

In 1985, a small group of living history re-enactors, including myself, decided to recreate the Independent Company and conduct a living history program. Being a natural dyer and having worked extensively with piece goods, I volunteered to dye the material for at least some of the uniforms, using methods to the best of my knowledge comparable to British dyehouses of the period. Research dictated madder-red wool breeches, waistcoat and regimental coat for Privates and Corporals, with "Popinger green" cuffs and facings on the coat. "Popinger green" was determined to be popinjay or parakeet green, a light yellow-green. Officers' uniforms were similar, except that cochineal scarlet was used instead of madder red. Also, the Drum Sergeant-Major's uniform called for reversal of the colors on the regimental coat, i.e., popinger green body with scarlet cuffs and facings, and a most handsome tall embroidered miter cap of green and scarlet instead of a black tricorne. Finally, heavy brown linen gaiters completed the outfits. Originally, these were painted with a mixture of linseed oil and burnt umber. Instead, I decided to dye mine dark Indian cutch brown to reduce stiffness and mess.

I elected to do my Private's uniform first. I had on hand proven good-quality madder roots from Alliance Imports for the red. The "popinger green" would have resulted from dyeing woolen cloth light indigo blue in a fermentation vat, alum-tartar mordanting the blue cloth and overdyeing with weld or old fustic yellow. Fortunately, I had natural indigo,

---

JIM LILES *is a Professor of Zoology at The University of Tennessee in Knoxville, whose research interests include traditional dye technology and history. His book* ***The Art and Craft of Natural Dyeing: Traditional Recipes for Modern Use*** *will be published by the University of Tennessee Press in late 1990.*

purchased by my wife from Skilbeck Bros. of London in 1964. This family firm served as purveyors of the finest natural dyestuffs in the world from 1650 to about 1970. I also had plenty of old fustic (a tropical dye wood).

The next step involved procuring suitable 100 percent woolen cloth, about 10 oz./yard. This is ideal weight for sewing, particularly since the entire outfit would be linen lined. Good quality woolen cloth costs $10-20 per yard, and 5 to 6-1/2 yards are required, depending upon the size of the man. Thus, a reasonable outlay would be required for the undyed wool alone. Such a situation dictates running test samples. I knew my dyestuffs and mordant chemicals to be good, but would a selected cloth dye evenly and without spots, and would it dye a nice madder red rather than a reddish-orange or brick color? Also, how much madder would be optimum, and how much mordant?

Dyeing the Cloth

Dyeing piece goods presents some problems compared to dyeing fleece or yarn. One factor is the volume of fluid needed. I allow 4-1/2 to 5 gallons per pound of fleece or yarn, but prefer at least 6 to 8 gallons per pound of fabric. (Otherwise, the dyeing is likely to be uneven.) The additional volume dictates a large pot for mordanting and dyeing; I have both 12-gallon and 30-gallon vessels. A 3 yard (equals 3 lbs.) piece of fabric is about as large as my dyeing vessels, heat source, washing sinks and drying table can accommodate. Also, compared to a skein of yarn, a 3-yard piece of dripping-wet woolen cloth, 58 inches wide, is a rather formidable object! It's all I can handle.

A 3-yard piece is sufficient for the breeches and waistcoat for an average-to large-sized man, but an additional 3-yard piece is necessary for the regimental coat and fatigue hat. (A fatigue hat was worn during work detail; with green rim, red body and red and green tassel, it somewhat resembles an elf's Christmas hat.) Therefore, scouring, mordanting and dyeing had to be done twice for each uniform, using methods as identical as possible in both cases — same batch and amount of mordant and madder, same dyeing time, etc. I dyed the cloth for other mens' uniforms in my 30-gallon stainless steel vessel, but made an exception for my own uniform. Since I am quite small, only 2 yards was needed for my regimental coat, 1 yard for the waistcoat, and 1-1/2 yards for the breeches and fatigue hat. I used my 18th-century 12-gallon copper dyeing vessel in this case, but had to do the madder dyeing three times. Some would consider this going beyond the call for strict adherence to authenticity.

By sampling, I determined that the best results were obtained by premordanting the woolen fabric with 4 oz. alum and 1 oz. cream of tartar per pound of wool, followed by dyeing with 8 oz. madder roots per pound of wool. During the dyeing, I manipulated the fabric almost continuously, repeatedly turning and unfolding it so that it would dye evenly. Piece goods always require much more handling than is the case with yarn or fleece. To prevent spotting, I am careful never to permit piece goods to cool in any dyebath.

Dyeing the popinger green proved to be rather difficult. At first I had beginner's luck with respect to depth of the indigo blue and strength of fustic yellow overdye, but it was hard to match the color for subsequent uniforms. One would wonder why a light yellow-green would have been used. Such greens are prone to fade with exposure to light, though I must confess that the color is quite attractive with either madder red or cochineal scarlet. Two possibilities come to mind. One is that more durable, darker greens may have already been in use by other regiments. The other is that the regimental

Miter cap — cochineal, scarlet and popinger green.

PHOTO BY DALE LILES

coat was replaced each year. The first year the soldier was outfitted completely; the second and subsequent years he received one new shirt, neckstock, and pair of shoes, as well as a new regimental coat and pair of breeches. The old regimental coat was then cut down into a replacement waistcoat, sleeved or sleeveless, but in any event the green facings and cuffs were discarded. I attempted to make the green a bit more fade-resistant by adding a small amount of copper sulphate after-mordant to the yellow dyebath near the end of the dyeing cycle. This precaution helps, and while I was not particularly worried about my own facings and cuffs, I was considerably concerned about the Drum Sergeant-Major's regimental coat, predominantly colored popinger green.

Dyeing the heavy linen for the gaiters cutch brown presented no problem. The tricks here are not to add the material to the hot cutch bath until all dyestuff is completely dissolved, and to finish the dyeing in the hot (170-190 degrees F) dyebath. Permitting dyed piece goods to cool in a cutch bath is particularly disastrous since catechin, one of the two dyes in cutch, is soluble only in hot water.

This concluded the dyeing for my Private's uniform. Dyeing cochineal scarlet for the officers' uniforms took a bit more effort than necessary as I adhered strictly to 1750's methods. This involved making the mordant (tin or red scarlet spirits) from feathered (powdered) tin, and a combination of hydrochloric and nitric acids and water. The dyeing was one-pot, using the tin spirits and tartar, along with 1 oz. of ground cochineal and a touch of turmeric per pound of wool.

Each member of our group practices a craft occupation, just as the original Independent Company did. One couple are leather workers. Others make our handsome cowbag knapsacks and linen haversacks. Our Drum Sergeant-Major is a pewterer, molding the many buttons for our uniforms, as well as spoons and other utensils. He also conducts Sunday morning services; quite appropriately, he is Episcopalian.

MAKING A TENNESSEE HERITAGE SAMPLER QUILT

JIM LILES

When I started researching pre-synthetic-era dyes, one of the first things that caught my attention was the fact that the majority of 18th and 19th century dye manuals contain as many or more recipes for cotton and linen as for wool or silk. By contrast, the natural dyeing of cotton and linen has been largely neglected in the 20th century. Therefore, I chose to concentrate my efforts on these fibers, using and modifying recipes I found in the old dye manuals, and I have been richly rewarded with beautiful colors. Although the processes are generally more involved, time consuming and expensive than with wool, cotton without question has become my favorite fiber to dye.

Cotton is the major fabric of quilters, and by 1984 most of my natural dyeing was being done on cotton sateen and print cloth in pieces 1 to 2-1/2 yards long. Increasingly, I wished to see my dyed fabric in a quilt rather than piled up in boxes, so I approached the "Thursday Bee" of the Smoky Mountain Quilters of Knoxville and Oak Ridge,

Tennessee, with a proposition. Would members of the bee be interested in making a quilt of my fabrics? The response was most gratifying, and within two months a plan was formulated to produce a Tennessee Heritage sampler quilt, similar to those of the period from 1830 to 1870. The restrictions were that the quilt block patterns had to be those found in Tennessee quilts made prior to 1930, but the blocks could be of varying size and of any color combination.

Our goals in this project were several. We wanted to show how such a quilt would have looked, newly made. A few of the dyes, such as indigo, Turkey red, black oak bark yellow, chrome green and antimony orange are remarkably lightfast and appear very much the same in an old quilt as in one newly made. Many others however, fade at least somewhat. The quilt would show that the natural dyes of our ancestors were not always sombre and muted, but also bright and vibrant, though with a distinct softness. Bright, vibrant colors were highly desired for dark cabins or houses illuminated by candlelight or coal oil lamps. It would also demonstrate that natural dyes, premixed by nature, present a different and often more pleasing visual effect than do many of our highly pure modern synthetic dyes.

The quilt illustrates the range of colors possible and in use by 1830. Beautifully colored fabrics were available to those who could afford them and those with the expertise to dye their own. By 1800, the use of the traditional cotton dyes had reached the peak of its perfection; by 1830 several additional mineral dyes were available and cochineal was in use by the calico printers. Thus, a full array of plant-derived, animal-derived and mineral dyes were in use.

A final goal for this project was to recreate the quiltmaking experience of our ancestors. We decided that the quilt would be entirely hand sewn, and my wife Dale would hand card the cotton batting, as in the originals. Working on the quilt gave us a much closer feeling and appreciation for the problems of our quiltmaking ancestors. For example, quilting through unscoured cotton batting proved to be a pleasure because the traces of cottonseed oil and wax serve as a natural lubricant. Conversely, quilting through dense mineral-dyed fabrics resulted in sore and pricked fingers, a frustration for all. Quilting stitches had to be close, because a quilt filled with traditional cotton batting becomes lumpy upon washing if the quilting stitches are farther apart than the short length of the cotton fibers.

In all I dyed approximately 120 colors, tints, shades and intensities; one square yard being the smallest quantity of each. Often the same dye can provide several colors or intensities of color. For example, Bemis (1815) listed 13 shades of indigo blue. Madder gives pink, red, purple, brownish-purple, gray, black and reddish-brown, depending on mordant combination and dyeing procedure. Madder with black oak bark yellow results in an array of oranges; madder with indigo gives purple.

I selected dyes exhibiting the greatest color range, while having the best reputations for light- and wash-fastness. I used no brazilwood or logwood reds since these fade to a rather ugly brown or pink-purple. Other rather fugitive dyes were used sparingly. Upon completion each block was placed on a wall in the workroom, illuminated by fluorescent lights, until the project was completed. Some of the first blocks were up for a period of two years, long enough to detect any fading dyes. By now the quilt has been under strong lights in four exhibits, a duration of five to six months, and fading has been minimal.

Dyes used in the quilt

Reds and pinks: Madder, madder and black oak bark, Turkey red, cochineal and coreopsis.

Oranges: Madder and black oak bark, coreopsis, chrome, iron buff.

Yellows: Black oak bark, goldenrod, chrome, iron buff.

Greens: Indigo and goldenrod, indigo and chrome yellow, Prussian blue and chrome yellow, Prussian blue and old fustic, logwood and old fustic.

Blues: Indigo, indigo and cochineal, indigo and iron tannate, Prussian blue, Prussian blue and logwood, Prussian blue and safflower, Amish logwood.

Purples: Madder, cochineal, logwood.

Browns: Bombay cutch, cutch and osage orange, osage orange and logwood, cutch and walnut, coreopsis and tannin, mineral khaki, manganese bronze.

Grays and blacks: Iron tannate, iron tannate and mountain laurel, iron tannate and logwood, Prussian blue and mountain laurel, logwood, madder and black oak bark. ❖

This sampler quilt is made of cotton using traditional dyes.

Photo by Dale Liles

Cochineal and indigo produce a wide range of colors. PHOTOS BY AUTHOR

USING COCHINEAL AND INDIGO IN DYE COMBINATIONS

TRUDY VAN STRALEN

I dye hundreds of pounds of wool, mohair and silk fibers and yarns each year with natural dyestuffs. To get the widest range of colors at a reasonable cost, I often combine expensive exotic dyestuffs such as cochineal and indigo with local dyes — any of the plants growing nearby which give yellows and browns. I'll share some of my recipes in this article.

I like to work by percentages so I use the metric system for weights and measures. Don't let this scare you — people who participate in my dye workshops soon adjust to metric measures, and you can too. Increased accuracy is another

TRUDY VAN STRALEN *lives in rural eastern Ontario, Canada. She has been working with natural dyes for 15 years, and sells vegetal-dyed fleece and yarn to handspinners and weavers.*

advantage to working this way. If you're careful about weighing the fiber and dyestuff on a gram balance, you can repeat recipes consistently. By contrast, if you work with teaspoons and cups you can easily be off by 10% to 20%, which makes noticeable differences in the results.

For each of the recipes below, be sure the fiber is warm and wet when adding it to a dyebath. After dyeing, wash the fiber and rinse it well before drying.

Using Cochineal

Cochineal alone gives a bright red. I also use it in combinations as a way to "warm" other colors. It's very expensive, but a little goes a long way. To get the most color out of every bug, I use a coffee mill to grind the cochineal into powder as fine as the finest melitta coffee.

No mordanting is required to make the color fast, but premordanting with alum or tin gives deeper colors and makes the supply of cochineal last longer. To mordant 100 gr. of fiber for cochineal and cochineal combinations, use 10% (=10 gr.) alum plus 5% (=5 gr.) tartaric acid; or 1/2% (=1/2 gr.) tin plus 5% (=5 gr.) tartaric acid. (Using the tartaric acid in combination with the alum or tin helps soften the yarn.) Mix the mordant with 4 litres of water, add the fiber and simmer for one hour.

1. BRIGHT RED FROM COCHINEAL. For 100 gr. tin-mordanted wool, use 10 gr. cochineal, and 5 gr. tartaric acid. Soak the cochineal and tartaric acid in a stainless steel or enamel dyepot overnight with 4 liters of water. In the morning bring the pot to a rolling boil and let boil for 15 minutes. There will be a foam on the surface that includes all the little particles of cochineal. Take some waste fleece and stir it around in the pot to catch all the residue. The dyebath is now ready for dyeing. Put the fiber into the pot and simmer for 30 minutes.

2. SOFT RASPBERRY FROM COCHINEAL AND BRAZILWOOD. For 100 gr. of wool, use 30 gr. brazilwood, and 3 gr. cochineal. Put the brazilwood chips in an old nylon stocking, combine with the cochineal in 4 liters of water and soak overnight. In the morning bring the pot to a boil and boil for one hour. Remove the brazilwood and use waste fleece to skim off the foam. Put the fiber into the pot and simmer for one hour.

3. WARM BROWN FROM COCHINEAL AND BLACK WALNUT. For 100 gr. of wool, use 500 gr. black walnut hulls, and 4 gr. cochineal. Cover the black walnut hulls with 4 liters of water and boil vigorously for 90 minutes. Strain off the liquid and add enough water to make a 4-liter dyebath. Add the cochineal and simmer an additional 15 minutes. Put the fiber into the pot and simmer for 90 minutes. The fiber will take on the cochineal color first, but needs the full 90 mintues to get the most out of the black walnut. This dyebath can be reused with an additional 100 gr. of wool for a lighter shade of brown.

4. WARM ORANGE FROM COCHINEAL AND MARIGOLD. For 100 gr. of wool, use 400 gr fresh marigolds, 3 gr. cochineal, and 5 gr. tartaric acid. Soak the marigolds overnight in 4 liters of water, and soak the cochineal and tartaric acid overnight in a small jar of water. In the morning, boil the marigolds for one hour then strain off the liquid. Add the cochineal solution and enough water to make a 4-liter dyebath. Boil for 15 minutes. Put the fiber into the pot and simmer for one hour.

Using Saxon Blue Indigo

This is not a traditional method of using indigo. Instead, it is an old industrial recipe, used in Europe just before chemical dyes were discovered. Compared to dyes from an indigo vat, this method gives a more royal blue color with neutral water and a more turquoise color with acid water. The recipe involves making a concentrated solution of dye liquid, which is then used a few drops at a time.

1. Making the Saxon blue solution. This should be done outdoors, with proper tools and protective clothing — a rubber apron, rubber gloves and safety goggles. Measure 14 gr. indigo and 7 gr. calcium carbonate and put them in a large glass jar. Carefully measure 90 gr. sulphuric acid into a separate container. Add a little of the sulphuric acid to the indigo and calcium carbonate while stirring with a glass rod. Repeat, adding just a little at a time. Then stir once a day for the next 14 days. After that the solution is ready to be used. Label the jar, fasten its lid securely and keep it out of the reach of children. This solution won't spoil and keeps for years.

2. Dyeing wool with Saxon blue. Add 1 ml to 3 ml indigo to 4 liters of warm water and stir. Add 100 gr. warm, wetted, unmordanted wool. Simmer for 30 minutes.

3. Aqua-turquoise from indigo and Queen-Anne's lace. For 100 gr. alum-mordanted wool, measure 200 gr. Queen-Anne's lace (or use another local plant which gives a yellow dye). Cover the plants with 4 liters of water and simmer for one hour. Strain off the dyebath, add the wool and simmer for 30 minutes. Lift wool out of bath and add 2 ml Saxon blue indigo solution. Stir. Put the wool back into the bath and simmer for another 30 minutes.

4. Forest green from indigo and *cutch*. For 100 gr. of wool, measure out 20 gr. of *cutch* extract and dissolve it in 4 liters of hot water. Add the wool and simmer for 45 minutes. Lift wool out of bath and add 4 ml of Saxon blue indigo solution. Stir. Put the wool back into the bath and simmer for another 45 minutes. This recipe can be used on unmordanted wool, or with different mordants to obtain different shades of green.

5. Purple from cochineal and indigo. Soak 5 gr. cochineal and 5 gr. tartaric acid in 4 liters of water overnight. In the morning heat and simmer for 30 minutes. Add 100 gr. tin-mordanted wool to dyebath and simmer for 15 minutes. Lift wool out of bath and add 3 ml indigo solution. Stir. Put the wool back into the bath and simmer for another 30 minutes. For brighter purple: Put the wool in a pail of hot water with 50 ml non-sudsy ammonia. Wash and rinse well.

6. Dark blue from cochineal and indigo. Prepare the cochineal dyebath as directed for purple, but use alum-mordanted wool. Using 3 ml of indigo solution gives a dark blue. For a lighter blue, use just 1-1/2 ml indigo. ✣

DYEING WITH WEEDS

ANNE BLISS

Natural dyers looking for reliable, cheap, readily available dyestuffs will find a treasure for their pots in vacant lots, along roadsides and in their own backyards. What those dyers will find are weeds that generously supply fast colors. Weeds belong to numerous plant groups, but they all share the characteristic of growing where they aren't wanted.

Weeds are quite successful at doing just what plants are supposed to do: adapt, grow and multiply. Humans don't have to plant their seeds, make cuttings, water or feed these plants to get them to thrive. Weeds do the work all by themselves, though many species seem to appreciate a farmer who breaks new ground or a homeowner who ignores a portion of the lot; it's in those open, idle grounds that weeds most often settle.

Weeds thrive around humans, or perhaps we should say that humans historically have made life easier for weeds. Ancient nomads, who moved frequently, transported seeds and dropped them along the way. We and our pets do the same thing today when we go hiking and come home with socks full of seeds. Early agriculturalists enabled certain plants to spread into broken or tilled soil; the disturbed ground, bared of native flora, allowed "immigrant" species to gain a foothold. In the mid-19th century, seed from Eurasia, imported for grain fields, contained errant weed seeds. As a result, a number of Eurasian "grain crop weeds" such as Russian thistle (*Salsola kali*) were introduced. These weeds have spread rapidly across North America and have naturalized in areas where they could adapt to soil and climatic conditions.

We can sort weeds into the 3 "N" categories: native, naturalized and noxious. Native species have been growing in the locale in which we find them since "who knows when." They came with the territory. Naturalized weeds are those introduced species that either were planted by mistake, as Russian thistle, or escaped from a purposeful planting. In general, naturalized weeds tend to interfere more in agriculture than native weeds and to create more problems as a result of that interference. The third category, noxious weeds, includes both native and naturalized plant species. Noxious weeds are so labeled by the U.S. Department of Agriculture because they severely deplete the soil of valuable nutrients, poi-

ANNE BLISS *is the author of* ***A Handbook of Dyes from Natural Materials*** *and a field guide entitled* ***North American Dye Plants***.

son animals, or generally cause extensive agricultural damage and economic loss. The law forbids growing, selling or propagating noxious weeds.

PHOTOGRAPH BY AUTHOR

Dyes from the weeds in the bouquet above were used to color the surrounding yarns.

Although only ten weedy plant species are discussed here, thousands more are available. Some weeds that produce wonderful, but not always fast, colors are localized in certain parts of North America. An example is pokeweed or pokeberry (*Phytolacca americana*), which produces lovely but fugitive cranberry-reds, violets and oranges. Weeds, like non-weedy species, have relatives that generally produce similar colors. If one species isn't available in your area, try a relative. Plants discussed here, however, are tough, common, widespread pests that can be found over most of the United States, Northern Mexico and Southern Canada.

If you aren't sure about identifying plants or knowing whether or not the species you find are weeds, order a copy of my favorite reference on North American weeds: *Common Weeds of the United States* published by the U.S. Department of Agriculture (1971) and available in paperback from Dover Publications, Inc. in New York.

The colors produced by the plants described here appear quite similar at first glance. No wonder ancient dyers got excited about indigo blue or the cochineal insect and its red dye; our ancestors also got the golds, yellows, and olive colors you see here from "local" plants. By manipulating picking time (before or after blooming, before or after rain) and time in the bath (1 hour more or less followed by cooling or fermenting for as long as you like), you can produce lighter colors, yellow instead of brass, and so forth. Experimentation and record keeping are keys to colors and success; bear in mind that water and soil conditions (acidic, alkaline, predominant minerals) can affect your color outcome.

These ten weeds were used in a simple pre-mordant and dye method that you can adapt to your own facilities and needs. Use rubber gloves, a face mask, and good laboratory procedures when working with mordants and dyes. Have good ventilation, and dispose of exhaust baths safely. Store mordants away from children and unsuspecting adults; label mordants as poisons. Use common sense in your weed gathering; skip those with known toxicity to humans (e.g., poison ivy) and those to which you're allergic (or use a ventilator mask that filters the air you breathe).

Here's my simple "test" method; you can vary amounts once you get a benchmark reading.

MORDANT: Per 1 pound of dry weight clean wool use 4 gallons of water in a non-reactive pot with lid. Add mordant choices to water and stir to dissolve:

Alum: Potassium aluminum sulphate: 4 tablespoons (64 grams)

Chrome: Potassium dichromate or bichromate: 3 teaspoons (20 grams)

Copper: Copper sulphate: 2 tablespoons (42 grams)

Tin: Stannous chloride: 2 teaspoons (12 grams)

Iron: Ferrous sulphate: 2 tablespoons (27 grams)

Enter well-wetted wool (soaked in like temperature water for 30 minutes); bring to simmer and maintain temperature at least 1 hour. (Optional: Cool yarn in mordant bath.)

Remove yarn and rinse 3 times (more if water is not clear).

DYE: Per 1 pound of dry weight clean wool use 4 gallons of water in a non-reactive pot with lid. Add 1 pound of fresh plant to water (1:1 plant to wool ratio); you can adjust the 1:1 ratio after a test (use more plant to get deeper colors, less for lighter). If you chop the plant, pigment acids will release faster into the water. Bring to simmer and maintain the temperature at least 1 hour. (Optional: Cool plant in bath overnight.) Strain out dyestuff. Add premordanted, rinsed wool (same temperature as dyebath to avoid shocking the wool) and bring to a simmer. Maintain the temperature for 1 hour. (Optional: Cool yarn in bath.) Drain; rinse yarn in clear water until water indicates no further bleeding; i.e., until the rinse water remains clear. Dry, under even tension, out of direct light.

Allowing wool or dyestuff to remain in cooling baths often produces better wash- and lightfastness than rinsing immediately after the 1 hour simmer. Note that colors tend to darken during the cooling period. The colors listed below were recorded after cooling. Each color was lightfastness-tested in bright Colorado sunlight (5,400 feet) for 50 hours, and fading or change in coloration was noted.

Bindweed

(*CONVOLVULUS ARVENSIS*)

This weed was introduced from Eurasia and became a bane of North American gardeners, but a delight to dyers who find the excellent fastness and color range easy to accept. Bindweed is a perennial, with spreading roots that may grow down 20 feet, as well as a bountiful seed crop following pretty white or pinkish morning-glory blossoms. Use as much of this noxious weed as you like, for by pulling on the vines, you'll break the roots, which in turn magically send up multiple shoots (more dye!). Flowering vines with leaves gave these colors in mid-summer: No mordant — light yellow; Alum — yellow; Chrome — bright gold; Copper — greenish gold; Tin — Yellow; Iron — olive.

Fastness: no visible fading.

Cocklebur

(*XANTHIUM ITALICUM* AND OTHER SPECIES)

This rough-to-the-touch, tough annual weed produces small yellowish flowers followed by hard brown seed burs that give the plant its name. After an introduction from Eurasia, cockleburs have spread from coast to coast, except for a small area from upstate New York east to the southern tip of Maine (Beware, they're headed your way!). The genus *Xanthium* contains several species, varying more in the characteristics of the burs than in dye colors. Some cocklebur plants grow to be 5 feet tall, so one plant may fill your dyepot. If you live in an area with livestock, make a lot of cocklebur dye because the seedlings are poisonous to the stock. Early August's greenish burs, leaves and stems gave these colors: No mordant — light brass; Alum — brass; Chrome — brass; Copper — light greenish gold; Tin — brass; Iron — dark olive green.

Fastness: no visible fading; iron turned slightly browner

Dogbane

(*APOCYNUM CANNABINUM*, *A. ANDROSAEMIFOLIUM*)

Dogbane is poisonous to cattle but has historic importance to several Native American groups who used the stem fibers for basketry; another common name for dogbane is Indian Hemp. Look for dogbane in damp soils, along roads and railroad tracks and at the edge of the forest. The spring flowers are tiny, greenish-white 5-petaled blossoms that develop into interesting, slim pods full of tiny parachute seeds. Although dogbane is a reliable colorist, be careful to wear gloves. The milky white fluid in the stems may irritate your skin. August's stems, leaves, flowers, and pods gave these results: No mordant — brass; Alum — bright brass; Chrome — brass; Copper — brassy khaki; Tin — brighter brass than alum; Iron — dark brown-green.

Fastness: very slight fading (alum, copper) and slight dulling (tin, iron); only chrome and no mordant showed no change. All were fast to washing.

Dock, Curly Dock
(*Rumex crispus*)

Another introduced weed from Eurasia, curly dock, has spread throughout the U.S., Canada and into Mexico. (Some other dock species are native to the U.S.) In some areas this 3-feet-tall, wavy-leafed plant with a large yellowish taproot is classed as noxious. Even though the oxalates contained in the leaves have been known to cause problems for grazing animals, many people consider the young leaves with their slightly sour taste to be superior potherbs. The rust-brown seeds have been widely used as a grain among various Native American groups. Green leaves, stems and early seeds produced these colors: No mordant — brass; Alum — brass; Chrome — brass; Copper — khaki; Tin — bright brass; Iron — dark olive.

Fastness: no visible fading; tin turned slightly pinkish.

Goldenrod
(*Solidago* spp.)

Approximately 100 species of this weed can be found in the U.S. and Canada. Some species grow in damp woodlands, others in dry open areas; related species tend to cross boundaries so more than one species can be found in some areas. Goldenrod, native to North America, is a cultivated horticultural plant as well as a bothersome weed, depending on where and when the plant grows. This plant has a long history as a source for yellow and gold dye; our colonial ancestors used it instead of imported weld and other dyes. The top half of the 5-feet-long stems with leaves and blossoms produced these colors: No mordant — yellow; Alum — bright yellow; Chrome — gold; Copper — bright green-gold; Tin — very bright gold; Iron — dark olive green.

Fastness: no visible fading from light or washing.

Mallow
(*Malva neglecta*)

Mallow is a somewhat deceptive weed introduced from Eurasia. The low growing, limp-stemmed plant has small white flowers followed by little seed wheels that give the plant one of its common names, "fairy cheeses." The inch-wide leaves that stay green all summer and the unripe seeds can be used as pot greens. Gardeners often dislike mallow, however, because even half-grown plants can be quite difficult to pull, and unless the root is cut off well below soil surface, it will send out new stems. Stems with blossoms, leaves and fruits gave these colors in mid-summer: No mordant — light brass; Alum — golden tan; Chrome — gold; Copper — gray-green; Tin — gold; Iron — dark olive.

Fastness: Iron browned and tin dulled; others showed no visible change.

Mullein
(*Verbascum thapsus*)

Here's another naturalized Eurasian plant that has spread across vacant land, especially dry rocky areas. Mullein is biennial, with a furry rosette of leaves the first year topped by a tall (to 6 feet or more) stalk with an elongated spike of small yellow flowers in the second year. Common names such as "Torchweed," "Hag Taper" and "Parson's Staff" refer to ages past when the dry stalks in winter were dipped in tallow and used as torches to repel witches. You don't have to be either a witch or a parson to get a fine set of colors from leaves and blooming stalks in August.

No mordant — gray-yellow; Alum — gray-yellow; Chrome — brass; Copper — brass; Tin — brass; Iron — dark olive.

Fastness: Iron browned slightly in light; others unchanged from light or washing.

Ragweed (Giant)

(AMBROSIA TRIFIDA)

You'll find one or another member of the ragweed clan in every state and across southern Canada, and they all make some people sneeze during pollen season. You can help cure the sneezing by brewing lots of ragweeds in your dyepot. Of the three most common ragweeds (*A. trifida*, *A. artemisiifolia* and *A. psilostachya*), only giant ragweed is a European import. This plant grows tall (10 feet or more in fertile soil) and has tiny spikes of yellow flowers that bleed red it you pinch them. Alas, not red dye, however; but you can get some really fast "composite family" shades at the same time you eliminate some pollen-making weeds.

No mordant — brass; Alum — yellow-gold; Chrome — rust; Copper — green gold; Tin — bright brass; Iron — dark olive.

Fastness: no visible fading.

Redroot Pigweed

(*AMARANTHUS RETROFLEXUS*)

Although a native of tropical America, this tough, adaptable plant has spread across the U.S. and southern Canada. In some areas redroot pigweed has poisoned cattle with its concentrated nitrates, but humans eat the dark green leaves as a potherb. Redroot pigweed, which you're as likely to find in your garden as in the "wild," has a red taproot, thus its common name From leaves, flowers and washed stems you should get colors like these:

No mordant — brass; Alum — light gold; Chrome — brass; Copper — brass; Tin — bright brass; Iron — olive.

Fastness: very slight fading in light and from washing.

(Canadian) Thistle

(*CIRSIUM ARVENSE*)

Finally there's a use for this prickly, noxious weed that blankets the southern part of Canada and the northern half of the U.S. Although it's called Canadian thistle, it comes from Eurasia like its cousin, the bull thistle (*Cirsium vulgare*). Thistles generally are wonderful colonizers; the Canadian thistle, a perennial, has deep, spreading roots and produces too many parachute seeds to count. Its growth habit allows it to rapidly fill waste areas and fallow cropland. Once it's settled, it's nearly impossible to eradicate. You can use all the stems, leaves and flowers you want of this plant and its relatives to produce some interesting colors. Though the colors fade slightly, the longer the fiber spends in the dye (e.g., cool 24 hours after simmering), the faster the dye. Don't let the wool ferment in the thistle bath, however.

No mordant — light gray yellow; Alum — light brass; Chrome — brass; Copper — brass; Tin — brass; Iron — olive.

Fastness: fades slightly in light; no visible fading when washed. ❖

PHOTOGRAPH BY RITA BUCHANAN

Wild plants collected from hills and fields can yield dyes.

WILD PLANT DYES

J. DANIEL SAFFER

As a boy I loved searching the fields and hills around my home in Frederick County, Maryland. Weeds and wildflowers fascinated me then, and years later, my interest was rekindled by the wonderful handbooks of the Brooklyn Botanic Garden. I used to work at the Monumental Printing Company in Baltimore, where the handbooks were printed. Indeed, I was one of the proofreaders who read the first dye handbook, *Dye Plants and Dyeing*, when it was prepared in 1964.

J. DANIEL SAFFER *is a retired typesetter who researches and collects native plants in the Baltimore area.*

Alas, the Monumental Company closed as offset printing replaced letterpress printing, and I faced an unwelcome early retirement. Looking for something to fill my time, I began to collect and study the wild plants of my area. For years I have recorded the first date of bloom for every weed and wildflower I see; in 1988 I had over a thousand different entries. I like to know the uses and lore of each plant.

Wild dye plants, with their history going back thousands of years, are especially interesting, but they can also be challenging and frustrating. Even a rank beginner in the fields and woods can find many dye plants that can be processed very easily. But instant gratification is not for the wild dye maker. Our subjects require time and patience. I soak plants overnight or for several days, sometimes weeks, in water with a pinch of detergent (never soap) to bring out the color. Patience must be exercised not only in the long soaking but more especially in the final dye preparation. Bring it to a slow simmer, avoiding intense heat. I have ruined many batches of dye by cooking them too fast and too hot, instead of simmering, then waiting. For the best, richest and deepest colors, use the lowest possible heat. Even the finest dyes revert to ugly tans and browns if they are overcooked.

Tans and browns constitute about nine-tenths of the wild plant dyes; you don't get many bright colors. I think that the best and brightest dyes come from the autumn berries. Red sumacs give a coral pink which shifts to lavender or dull purple with the addition of a few rusty nails. Privet berries usually give a nice gray or blue-gray. Fruit of the viburnums, especially maple-leaf viburnum, give blue-violets and purples. Wild grapes give deep lavenders and the fruit of greenbrier or smilax usually gives a blue-gray. Fruits of Boston ivy and its close relative Virginia creeper also give fabulous dyes in the blue-gray-purple range, and their bright red autumnal leaves give a nice red-brown, although not the red that we would desire.

Never count on the shades and colors you are told a plant will make. They vary by any number of factors, and each time you may get a different result. From well-known authors I read of a fabulous vivid orange dye from nutgrass (*Cyperus*), but it never gave me anything like orange. Sometimes, though, I stumbled upon greater dyes than I read of in any book. Once I noticed a mass of tree-of-heaven (*Ailanthus*) leaves in a puddle after a long rain. They had a certain glow and made a fabulous bright ocher dye.

A good way to see if the plants in question will produce a dye is to rub it on a sheet of white paper. This color will usually be brighter or darker than the actual dye, but it gives you a good idea. My first rubbings happened by accident when I noticed that the specimens of jewelweed I gathered for my herbarium produced a bright orange color against the pages of my collecting book.

I don't dye yarn or cloth; I dye paper. I make labels for each common wild plant of the Baltimore area, and when I process a natural dye I dip a set of labels in the dye which that plant makes. In the winter of my retirement, I get comfort from this hobby and the course to which I was steered by the handbooks of the Brooklyn Botanic Garden many years ago.

A Black Dye from Purple Loosestrife

Su Grierson

Purple loosestrife (*Lythrum salicaria*) has a lovely Gaelic name: Lus-an-sith-chainnt, or "peace-making" plant. It grows in ditches and wet areas, and makes quite a show in July and August when the strong upright stems are heavily loaded with spikes of deep purple flowers. Although it is an uncommon plant in Scotland, in some areas it grows abundantly and and it is a beautiful plant to add to a dye garden. (Editor's note: Gardeners in the United States are advised not to plant purple loosestrife, which has become an aggressive weed in wetlands along the East Coast and the northern states. Dyers in those areas should take advantage of existing stands.)

For use as a dye, the flowering tops and stems should be gathered, chopped to fit the dyepot, just covered with water and then boiled until the liquid is the color of strong tea. Allow 40 minutes for this, but longer boiling will do no harm. All dye plants are variable in tinctorial power, but purple loosestrife generally has good dye strength. The plant is rich in tannin. Test a 2:1 ratio (by weight) of fresh plant to dry fiber to start with, then alter the ratio as needed to achieve the desired depth of color.

Strain off the dye liquid and add unmordanted wool. Soak the wool in the dyebath overnight and reheat to the simmering point the next day. Remove the wool while adding just enough ferrous sulphate to turn the dyebath a good black color. Then put the wool back in the dyebath and simmer it, always below the boiling point, until a deep black dye is obtained. Depending on the nature of the fiber, it will take at least an hour to achieve this color. It can be left for longer.

WELD

HETTY M. WICKENS

Weld is one of my favorite dye plants. It grows well in my garden, the nodding spikes of flowers seem cheerful, and above all, the yellow and gold colors obtained from every aboveground part of the plant are most attractive and reliable. Weld has a long history. The Romans used it to dye wedding garments and the robes of the Vestal Virgins — the maidens chosen to tend the sacred fire in the temple of Vesta, Goddess of the Hearth.

Also known as dyer's weed, dyer's rocket and wild mignonette, weld (*Reseda luteola*) grows wild throughout Europe, and has been cultivated in France, Italy, Germany and Belgium. In Britain it grows on waste ground – on the chalk Downs, on roadsides, especially near new road works, and on some pit heaps in West Yorkshire.

From seed it grows a distinctive rosette of long narrow green leaves, with a long thin taproot. A long smooth stem grows up from the rosette, first as a single spike of tiny yellow and green flowers, later branching to bear side spikes of flowers. Weld flowers are heliotropic; in the morning they face east, gradually following the sun until by evening they have turned to the west. Normally weld grows to about 1 meter (just over three feet) in height, but the plants in my garden often reach two meters. Some dyers prefer to use a shorter weld, but so far, my tall weld has been very satisfactory. One of my plants had 36 spikes branching off the main stalk, and I counted at least 300 flowers on one spike. Each flower bears about 25 tiny seeds, so one spike could contain 7,500 seeds; the whole plant could make 250,000 seeds!

Thinking of seeds leads on to the subject of harvesting. Collect weld in June or

HETTY M. WICKENS *lives in Ashington, West Sussex, England. She is the author of* ***Natural Dyes for Spinners and Weavers*** *(London: Batsford, 1983), and before retiring, she was senior lecturer in the textile department at Coventry College of Education, now part of Warwick University.*

July, just before the seeds are ready to scatter. The lower leaves will have started to turn yellow by this time. The little black seeds are easy to see, packed tightly inside the tiny green and yellow capsules which followed the flowers. Take care not to scatter the seeds while cutting the weld, unless numerous plants are required. If weld has taken a liking to your garden, it could become a nuisance, with little rosettes of bright green appearing in unexpected places — weld plants are thriving now in my strawberry tub. However, I've found that the seeds don't scatter unless I've left the harvesting until too late. At the right time, they seem to be firmly fixed to the stems. And it's easy to pull out or hoe up unwanted seedlings.

PHOTOGRAPH BY AUTHOR

Weld can be used fresh or dried.

Weld can be used fresh or dried, but I get a slightly stronger color from the freshly picked plants. To store it for later use, let the weld dry in the air and then tie it up in bundles. It will look like straw. When I dry weld, I place the cut stems loosely in a paper bag in order to collect the seeds, which are a valuable additional source of color.

Dyeing with Weld

The amount of weld used in the dyebath will vary according to the depth of color required and the quality of the plants being used. An average ratio would be a pound of weld for a pound of wool. Break or cut up the stems, spikes and leaves. (I did a few experiments to test for dye in parts of the plant that haven't traditionally been used; the seeds gave a clear yellow, but the roots produced no color.) Cover the weld with soft water and bring the pot slowly to a boil. Gradually the plant parts will soften and the water will turn yellow. Opinions vary as to whether the dyestuff should be left in with the yarn; if so, the color will be deeper than if the liquid is strained off. Add premordanted yarn to the dyebath and simmer for about an hour. Remove the yarn, rinse it in soapy and then in clear water, and dry it in the shade.

Weld has given me excellent results on wool, silk and nylon. Colors vary from lemon yellow or bright yellow to gold or olive, depending on the quality of the weld and the season when it was grown, the type of material being dyed and the mordant used. Alum brings out a clear lemon yellow, chrome gives a warmer yellow, tin brightens and iron turns the color to olive. A bit of chalk added to the dyebath makes the color more intense. When experimenting, have plenty of little skeins mordanted and ready to dip into dyebaths, so you can continue until the color is exhausted. Finally, weld is extremely useful in conjunction with indigo for producing a variety of greens. If dyeing with indigo first, try to avoid too dark a blue, as the yellow is less effective on a dark background.

FLOWERS FOR DYEING

Rita Buchanan

Many flowers in the composite family can be used as dye sources. Some are wildflowers, and others are popular garden flowers. A delightful way to obtain cheerful, sunny shades of rusty red, orange, gold and yellow is to grow rows and beds of coreopsis, cosmos, rudbeckias, dahlias, zinnias, marigolds and other composites. You can enjoy a beautiful garden, cut plenty of bouquets and still have enough left over for dyeing. All of these can be raised from seed at little expense and will bloom over a long season.

Plants to Grow

There are several kinds of coreopsis or tickseed, both annual and perennial, including one species known as dyer's coreopsis (*Coreopsis tinctoria*). Some are conspicuous wildflowers in parts of North America, and all are desirable garden plants with abundant daisylike flowers in shades of gold, rusty brown and mahogany red. The perennial kinds form clumps about a foot in diameter, and have long-stemmed flowers. The annuals usually form stemmy plants about two feet tall. All kinds of coreopsis bloom over a span of several weeks, and you can harvest flowers repeatedly during that season. The darker-colored flowers give rusty red and bronze dyes.

Cosmos (*Cosmos sulphureus*) 'Diablo', 'Sunny Red', or 'Bright Lights', are popular garden annuals that bear bright red-orange flowers in abundance over a long season. (The tall, old-fashioned garden cosmos with pink, rose or white flowers is *Cosmos bipinnatus*, and isn't useful for dyeing.) Sow the seeds in late spring, and thin to six inches to 12 inches apart. They will begin to bloom in about eight to ten weeks and continue until hard frost. For dyeing, cut or snap off the flower heads at full bloom or just as the petals begin to wither. The plants will bear more flowers over a longer season if you keep picking them regularly. Even if you pick the plants completely bare, more flowers will open in a day or two.

Black-eyed susans (*Rudbeckia hirta* and other species), dahlias (*Dahlia pinnata*), zinnias (*Zinnia elegans*) and marigolds (*Tagetes erecta* and *T. patula*) are familiar and popular among all gardeners. Flowers of any varieties of these plants can be used for dyeing, so plant whatever kind and color you like.

The pungent ferny foliage and the buttonlike yellow flowers of tansy (*Tanacetum vulgare*) give shades of yellow and yellow-green. The leaves and flowers of goldenrod (*Solidago* spp.) also give good yellows and greenish yellows. Gather tansy or goldenrod flowers just as they open for the clearest yellow dyes.

Dyer's chamomile, or golden marguerite (*Anthemis tinctoria*), is a bushy herbaceous perennial with ferny dissected foliage and daisylike yellow flowers that give yellow and tan dyes. Other chamomiles can also be used for dyeing, including the herb chamomile (*A. nobilis*) which make a soothing hot tea, and the weedy fetid chamomile, stinkweed or dog fennel (*A. cotula*).

Safflower (*Carthamus tinctorius*) has long been cultivated as a dye plant in China, Japan and India. An annual that grows quickly from seed, safflower looks like a thistle, with spiny leaves and stems and spiny-rimmed flower heads. The reddish-gold florets which form a tuft in the center of the flower head contain two distinct pigments, a yellow and a red. Both can be used for dyeing, but the red has always been more highly valued. Long ago, safflower red was used to dye cotton tapes for tying legal documents together — the original "red tape." Safflower is easy to grow, but each plant yields only a handful of florets, and it's hard to pluck them without running into the spines.

Dyeing with Flowers

Start with about two pounds of flowers in three gallons of water and simmer for a half hour or until the flowers are bleached, then strain off the liquid for a dyebath. This is enough to dye four ounces of pre-mordanted wool if you want dark, rich colors, or more wool if you want pastel colors. Put the wool in the dyebath and simmer for an hour, then either cool and rinse promptly for light colors or leave to sit overnight for deeper colors. If the first batch of yarn comes out dark-colored, try dyeing successive skeins for progressively lighter shades of the same hue. In my experience, dyes from flowers are moderately fast to light and washing on wool, but ineffective for cotton or other plant fibers. Quite a range of colors can be produced from these flowers, depending on how concentrated the dyebath is, what mordants and additives are used and how long the fibers are left to simmer or soak. You'll want to experiment with all these variables. In particular, these are good dyes for comparing the effects of different mordants. Also, check the effects of raising or lowering the dyebath pH by adding a few spoonfuls of baking soda or vinegar.

LADY'S BEDSTRAW
A SCOTTISH DYEPLANT FOR WOOL

SU GRIERSON

PHOTOGRAPH BY AUTHOR

An array of yarns dyed with natural materials.

Dyers in every country and during every age have sought red dyes of lasting value, and in every land at least one native red dye has usually been found. In Scotland, the major red dye was extracted from the roots of lady's bedstraw (*Galium verum*), a member of the madder family (Rubiaceae).

This perennial plant grows throughout Scotland in roadside verges and hedgebanks, and in dry sandy places, particularly coastal regions, where large colonies can form. It is easily cultivated and will produce a large root system in light soil. In the 19th century some attempts were made at growing lady's bedstraw commercially, as madder was grown, but it seems that none of these experiments were economically successful.

Lady's bedstraw has creeping, hairless, four-angled stems with many whorls of tiny single-veined leaves that are smooth

SU GRIERSON *is a dyer from Scotland.*

and dark green on top and downy below. Terminal panicles of small bright yellow flowers are usually borne on erect stems between June and September. The flowering tops of the plant can be used with an alum mordant to produce a yellow dye, but among Scottish plants, ling heather tops and bog myrtle leaves offer stronger, more permanent yellow dyes. Traditionally, using lady's bedstraw tops to coagulate milk for cheesemaking was more beneficial than using them for a dyestuff. Below ground, there are shallow rhizomes up to 1-1/2 inches thick and much finer roots which may extend six feet in dry soil. The red dye is present in all sizes of roots, but the color from the rhizomes is darker (less yellow) and more concentrated.

Although lady's bedstraw is a widespread plant in Scotland, it's difficult if not impossible to collect the roots in areas where a variety of vegetation has produced complex intertwined root systems. However, it is easy to extract long bedstraw roots from the grasses and flowers growing on the Machair — the sandy coastal areas that run along the western sea front of the Outer Hebridean Islands. Not surprisingly, the tradition of making red bedstraw dye was concentrated in these areas.

In the 18th and 19th centuries, visitors to the Outer Islands reported on the use of the red dye, but clearly it was a tradition with much earlier beginnings. Indeed, digging these roots, together with roots of tormentilla (*Potentilla erecta*) for tanning purposes, caused such unsightly scars and erosion that a Soils Preservation Act was passed in 1695 to prohibit the pulling of plants along the sea coast. But the desire for this beautiful red dye ran deep, and the local people continued to dig the roots on illegal expeditions in the dark of night. The islanders invested their clandestine escapades with their Celtic love of story telling and many a tale was told of uncanny nighttime happenings while digging the precious roots.

As with all dye plants, the tinctorial properties of the roots vary. Traditional gatherers — people from small communities who collected dye plants from the same limited geographical area for generations — learned that plants from specific sites or in particular growing conditions offered better or worse dyes, and they gathered them accordingly. This was clearly expressed by Winifred Shand (*Plants and Gardens* Vol. 20, No. 3). She reported that dyers on the Hebridean island of South Uist found one sack of local bedstraw roots was sufficient for a dyebath, but if they went to the island of Barra, less then 25 miles away, they needed to gather two sacks for the same results.

Other members of the genus *Galium* have historically been used for dyeing, but less so than lady's bedstraw. They possess a similar coloring matter, but their root systems tend to be smaller, and they often grow in situations where digging would be a difficult and unrewarding task.

Both northern bedstraw (*G. boreale*) and hedge bedstraw (*G. mollugo*) roots have been recorded as red dyestuffs in various parts of Europe, but they were not sufficiently plentiful to use in Scotland. Sweet woodruff (*G. odoratum*, formerly *Asperula odorata*), which grows wild in a few areas of Scotland and can easily be grown in gardens, produces a good red dye from its roots. Sweet woodruff is sometimes confused with dyer's woodruff (*A. tinctoria*), which grows wild in France and Denmark and was at one time used as substitute for wild madder. Wild madder itself (*Rubia peregrina*) does not grow in Scotland, but its use has been recorded in Ireland, where it grows on the Burren and Aran Islands.

For the modern craft dyer the most useful species, second to lady's bedstraw, is likely to be *G. aparine*, known by many local names but most frequently called goosegrass or cleavers. This plant with its creeping habits and millions of tiny clinging burs is a scourge of gardens and hedgerows. However, if dug out with much patience — essential anyway if it is ever to be eliminated from the garden — then a good supply of the very small roots can be obtained. Gathered in August and dried in the sun, these roots can produce red and rose colors on alum-mordanted wool.

How to Use Lady's Bedstraw for Dyeing Wool

I can't state the best age at which plants should be dug for dye purposes, but it appears that older plants yield more dye, and of a darker color, than young plants do. Also, the dye made from fresh roots is generally more orange in color than that from dried roots. Roots for dyeing can be dug at any time, cleaned and air dried, and chopped into lengths suitable for storage or use. The dye is thought to reside in the outer bark of the root and some traditional recipes suggest that it should be separated from the roots for dyeing, but good colors can be achieved without embarking on this laborious task.

Try starting with a 3:1 ratio of roots to wool (measuring dry weight). Depending on the quality of the roots and the depth of color required, you can amend this ratio for subsequent dyeings. A successful dyer doesn't rely on fixed recipes, but learns to make judgments and adjustments as needed.

Traditionally, roots were boiled for many hours before being strained, then mordanted wool was added to the liquid and boiled again during the dyeing. This traditional method is simple, but it often gives the wool a brick-red or brown-red color, due to the influence of yellow and brown pigments that accompany the red dye in the roots. During the 20th century in the Outer Isles, tin was sometimes used as a mordant for this dye, because it produces a more vivid red color on the yarn, even from a brownish dyebath. However, the tin-mordanted color is not lightfast, and becomes more yellow as it fades. Alum-mordanted wool fades less than that treated with tin, and turns pinkish instead of yellowish. Iron or chrome mordants give darker brown or purple shades.

Following a more complicated procedure gives clear red colors by reducing the uptake of the yellow or brown pigments: First chop the dried roots into small pieces, then soak them in cold water for 24 hours. Slowly heat the roots and soaking liquid and simmer at just below the boiling point until the dye liquid shows a good red color. Strain off this liquid into a second dye vessel and allow it to cool slightly. Add wet, alum-mordanted yarn and keep the second pot hot, but always below the boiling point. (This is important because apart from the possibility of damaging or felting the wool, higher temperatures bring out more of the yellow and brown dye components.)

Meanwhile, cover the roots with fresh water and slowly heat them again until the liquid is a good red color, a process that takes about an hour. By this time, the yarn will have absorbed most of the red dye from the first batch of liquid, which can be discarded. Strain the second batch of liquid off the roots and put the yarn into it for a second dyeing. This process can be repeated several times, stopping when the roots give only a brown color to the water. This method produces a rich red color on wool.

NEW WAYS TO USE EUCALYPT DYES

JEAN CARMAN

Since 1968 I have investigated the dyes from over 200 species of *Eucalyptus* from all parts of Australia. At first I primarily dyed wool, but in recent years I have used the eucalypt dyes to paint and dye silk and cotton fabrics and embroidery threads, and as a paste for screen and block printing. Here are three new methods for using eucalypt dyes. (Editor's note: These techniques may be adapted for using dyes from other plants.)

1. PRINTING PASTES FOR SCREEN PRINTING AND BLOCK PRINTING. Prepare a very strong eucalypt dyebath and reduce it to a thick paste by boiling or standing it in the hot sun. Add the paste to a printing base (available from textile crafts suppliers). The paste can be used with standard printing techniques on fabric or paper. For special effects, place pressed leaves or small flowers between the screen and the fabric or paper before printing. Fix the dye color by ironing the fabric 24 hours after printing.

2. THICKENED DYE SOLUTIONS FOR PAINTING. Use dye thickener (from crafts suppliers) to thicken a solution of eucalypt dye. The thickened dye can be used for painting on paper or clothing. Again fix the dye by ironing after 24 hours. The eucalypt dyes have proven color fastness and can be washed.

3. EUCALYPT SAWDUST PASTE FOR BLOCK PRINTING. Mix sawdust with printing base and apply to a wooden printing block with a paint brush. Press paper firmly against the block, and gently lift it off. Sawdusts from different species will give varied hues. Adding a bit of ferrous sulphate (iron mordant) to the printing base will give a gray-black color, and copper sulphate will give a greenish tone.

JEAN CARMAN *is the author of* ***Dyemaking with Eucalypts****, published in 1985 by Kangaroo Press. She lives in Queensland, Australia.*

EUCALYPT DYES ABROAD

During 1977 I visited China with a delegation of the Australia/China Friend ship Society. We travelled to Guangxi Province in southeastern China. Eucalypts have been planted extensively throughout China and my task was to introduce the eucalypt dyes to the textile people in Guangxi. They were using the eucalypt leaves as insect repellents and for medicinal purposes, and were interested to learn about their dye potential.

In 1988 I visited the Japanese Pavilion at Expo '88 (the celebration of the Australian Bicentennial), to meet with people from Saitama Prefecture in Japan, the sister-state of Queensland, Australia. It was to a zoo in Saitama that koalas were first sent in 1984; now koalas are kept in six zoos in Japan, and 33 different species of eucalypts have been planted throughout Japan to provide food for the koalas.

I learned that scraps of eucalyptus leaves which the koalas had not eaten were sent to the Textile Testing Station in Saitama, and dyes were extracted from them. The Japanese thought that this was the first time eucalypt leaves had been used for dyeing. They presented me with a silk tie which was finely woven with irregular stripes in muted shades of yellow, gray and gray-green.

We compared techniques of dyeing. The Japanese method for extracting dyes from plants, which has developed over thousands of years, is more complicated than mine. My method is simple and the colors depend very much on the species of eucalypt being used. One of the Japanese expressed his regret that the colors they had obtained were so muted. I told him that bright colors can be obtained, and gave the Japanese leaves from *E. cinerea* and *E. microcarpa*, which give bright red dyes. I also gave them samples of dyed wool and silk, and showed them beautiful examples of eucalypt dyeing done by other Australian craft workers.

A friend who recently returned from a trip to Japan told me that she visited the zoo in Saitama. In the gift shop there she saw a number of articles woven from silk dyed with eucalypt leaves, such as mats, handkerchiefs, scarves, ties and tablecloths with a motif of a koala in each corner. She told me these items were beautifully packaged — in some, a eucalypt leaf was included, and they were quite expensive.

— *Jean Carman*

DYEMAKING WITH EUCALYPTS IN AUSTRALIA

Diana Cockburn & Mary Jost

for the Plant Craft Cottage Dye Group

The genus *Eucalyptus* consists of over 600 species. They are all native to Australia and Papua, New Guinea, but because they tolerate different growing conditions, eucalypts now are planted throughout the world. All eucalypt leaves give color, but the excitement is in the range from pale yellow to deep dark red — mellow shades that reflect the unique colors of the Australian bush. The bark and fruit usually give more muted reds, grays and browns.

Many Australian craft workers use the eucalypt dyes. Our dye group has investigated them in a methodical way, and prepared a collection of dye samples for the Royal Botanic Gardens in Melbourne. Each sample is labelled with relevant details of the source, collection date and dyeing methods used. One series of dye samples is accompanied by herbarium specimens from the actual trees used.

There has always been discussion on whether fresh or dried leaves make the better dyestuff. The group experimented by stripping fresh leaves from a branch, mixing them well and weighing out many portions of equal weight. One portion was used as dyestuff while fresh; the others were allowed to dry and used after one week, then monthly for one year. Only slight color change resulted.

The intensity of color and the shade are affected by growing conditions and weather. Colors in years of flood or drought can vary dramatically from normal. To investigate the effects of season, we used leaves from two trees as dyestuff, picked at quarterly intervals for one year and then approximately monthly for two years. Color variations were evident, but no conclusions could be drawn because there were too many variables.

Dyeing Procedures

For each sample, we record the botanical and common name of the plant, the dates of collection and of dyeing, the part of the plant used, the weight of dyestuff, the type and weight of fiber and the recipe, including details of mordanting.

We normally use a 4:1 ratio of dyestuff to wool, although good colors can be achieved with less dyestuff. We cover the dried, crushed dyestuff with boiling water and let it stand overnight. Then we slowly heat the dyebath to 90 degrees C

Diana Cockburn and Mary Jost *are active in the world of textile crafts in Australia. Members of Plant Craft Cottage Dye Group have been studying eucalypt dyes since 1981 at the Royal Botanic Gardens, Melbourne, Australia.*

(195 degrees F), simmer at that temperature for one hour, and let it cool overnight. We use the following two methods of testing the color range obtainable.

Substantive Time Series: One of the most interesting areas of experimentation is a substantive time series. No mordants are used. Four skeins of scoured wool yarn are entered into the prepared dyebath and simmered along with the eucalypt leaves, bark or fruit. The skeins are removed one at a time after one, two, three and four hours of simmering. Colors obtained vary among species. Some give fawn shades only, others give oranges, tans, browns, reds, and dark claret after four hours' simmering. The dyeing may be streaky at first, but longer-cooked samples develop more even color. Boiling the dyebath with wool and dyestuff in it can produce ever darker colors, but the wool quality will suffer. The lightfastness of the substantive dyes is fair to good. Species from which dark reds have been obtained by this method include *E. cinerea*, *E. crenulata*, *E. melliodora*, *E. nicholii*, *E. polyanthemos*, and *E. sideroxylon*.

Gerber Investigative Method: This method, developed by American dyer Fred Gerber, produces 16 colors from one dyestuff. To start, eight skeins of wool are premordanted with alum and eight with chrome. Then portions of the prepared, strained, dyebath are modified with tin, copper, iron and ammonia. Mordant quantities per 500 gr. (slightly over a pound) of clean dry wool are 85 gr. of alum and 14 gr. each of chrome, tin, copper or iron. Weigh your wool first and measure the appropriate amount of mordant.

Here is the step-by-step procedure for dyeing alum-mordanted skeins. Follow the same sequence for the chrome series.

1. Put the eight alum-mordanted skeins into the dyepot. Heat slowly and simmer at 90 degrees C (195 degrees F) for one hour. Remove all skeins. Wash and rinse one skein, and label it "Alum a." Put a small amount of dyebath in a separate pot, add a teaspoon of household ammonia and stir well. Re-enter one skein, simmer 5 minutes, then wash and rinse it well. Label it "Alum b." Divide the remaining dyebath into three parts.

2. To one part, add tin mordant and stir well. Re-enter two dyed skeins and simmer for 15 minutes. Wash and rinse one skein and label it "Alum c." Add a teaspoon of ammonia to dyebath, re-enter the other skein and simmer for 5 minutes, then wash and rinse it well. Label it "Alum d."

3. To another part, add copper mordant and repeat the steps. Label "Alum e" and "Alum f."

4. To the last part, add iron mordant and repeat the steps. Label "Alum g" and "Alum h."

The lightfastness of the alum series is good, and that of the chrome series is very good. Different species give this range of colors: Green: *E. bicostata*, *E. callophylla*, *E. globulus*, *E. microcarpa*; Yellow: *E. cladocalyx*, *E. radiata*, *E. scoparia*; Yellow-Gold: *E. moluccana*, *E. sideroxylon*; Orange: *E. cinerea*, *E. crenulata*; Brown: *E. bancroftii*, *E. lehmannii*. These related colors in shades of yellows, greens, oranges and golds make an interesting resource for tapestry, needlepoint and other crafts. ❖

FURTHER READING:

Carman, Jean. *Dyemaking with Eucalypts.* Australia: Kangaroo Press, 1985.

Glasson, Ian and Mikki. *A Eucalypt Dyer's Handbook.* Stanfield, Carcoar, NSW: self-published.

Handweavers and Spinners Guild of Victoria. *Dyemaking with Australian Flora.* Australia: Rigby, 1974.

TRADITIONAL PLANT DYES IN JAPAN

SEIKO HIROTA

The Japanese people have enjoyed dyeing since ancient times. The original techniques were introduced from China or Korea, but by the 8th century characteristic Japanese techniques had started to develop. Dyeing skills improved over time and flourished from the 17th to 19th century; in that era, professional dyers called *chonin* achieved marvelous results. In Japan, two colors in particular — the blue from indigo and the red from safflower, are most important, and have been used since ancient times.

During the 20th century the old dyeing skills using natural resources have been discarded little by little, but we still have natural plant dyers in Japan. For example, Mr. Akira Yamazaki researched and published a book about the *kusaki-zome* or traditional Japanese plant dyes in 1961. Now his son Seiju Yamazaki is following his father's footsteps. In villages and towns you sometimes come upon exhibitions showing the work of amateur dyers who learned the craft at adult-education classes or at art colleges.

SEIKO HIROTA *is the author of two books on growing and using herbs. She lives in Kanagawa, Japan.*

Also we have an extremely special elementary school. It is located on Awaji Island between Honshu and Shikoku. The school's name is *shitoori*, which means "closely connected with cloth" in Japanese. All the pupils cultivate dye plants such as Japanese indigo (*Polygonum tinctorium*), safflower (*Carthamus tinctorius*), French marigold (*Tagetes patula*) and kusagi (*Clerodendron trichotomum*). Using their own plants, they learn dyeing techniques such as batik, *itajime* — a sort of variegated effect made with wood chips, and *tataki-zome* — a way of mashing the plants on the cloth. According to the pupils' age they learn to spin yarn and weave cloth. I think such artistic and historical lessons help us appreciate nature's blessings in our daily life, and it is likely that many schools will introduce this kind of curriculum in the future.

A Japanese Procedure for Dyeing Wool or Silk

1. Weigh the yarn or cloth, and weigh out the same amount of dye plants.

2. Put the dye plants into a stainless steel or enamel pan and cover with water. Bring to a boil and simmer for 30 minutes, then strain off the liquid. This is the dyebath. Let it cool down.

3. Gently wash the yarn in tepid water with a small amount of detergent. Rinse it well.

4. Put the wet yarn into the dyebath. Over a period of about 30 minutes, raise the temperature to 80 degrees C (180 degrees F). Don't let it boil. Simmer for 30 minutes, stirring with chopsticks from time to time to prevent uneven dyeing.

5. Meanwhile prepare a mordant solution. Dissolve an amount of alum equal to 10% of the weight of the yarn in enough water to thoroughly cover the yarn. Add the dyed yarn to the mordant solution, and slowly heat it to 80 degrees C for 30 minutes.

6. After mordanting, rinse the yarn, then return it to the dye solution and reheat it to 80 degrees C for a second dyeing. This fixes the color.

7. Wash the yarn thoroughly, squeeze out extra moisture, and let it dry in the shade.

A Sampler of Dye Plants

I find great pleasure in dyeing fabrics and yarns with wild plants and plants I have grown. How wonderful natural colors are! I will describe some dye plants which are familiar in Japanese everyday life.

HAMANASU or rugosa rose (*Rosa rugosa*) grows wild in the northeast Japan. This rose is a very hardy deciduous shrub with fragrant flowers and large rose hips which are used in preserves and medicines. The leaves, stems and rhizomes are used for dyeing, giving brown with chrome and reddish gray with iron.

KANAMUGURA or hops (*Humulus japonicus*) is a vigorous vine whose palmate leaves and stems are covered with irritating hairs, but the colors of hop dyes are very mild and subtle. Gather fresh leaves and stems in October. Hops gives pink with no mordant, beautiful yellow with alum, greenish-gray with iron and yellow-brown with copper.

UKON or turmeric (*Curcuma longa*) is a tropical plant from India and Indonesia that is cultivated in parts of Japan where the climate is mild. It has large leaves, beautiful white flowers and a yellow rhizome which is useful as a medicine and in oriental cooking as well as for dyeing. You can use the rhizome either fresh or dried. It gives a beautiful golden color with alum, golden-brown with iron and orange with tin.

CHA or tea (*Thea sinensis*) is an evergreen tree from China which can be grown in mild regions of Japan. It bears white flowers from October to December. For dyeing you can use green tea, black tea or Chinese tea. If you use the fresh green leaves, add a little wood-ash solution to make the dyebath alkaline. Tea leaves dye light brown with no mordant, light pinkish-brown with tin and gray with iron.

YOMOGI or artemisia (*Artemisia princeps*) grows everywhere along the banks of rivers. We eat the young leaves and chop the stems for a bath herb and a medicinal herb that aids digestion. This plant gives brown with alum, olive green with copper and dark greenish-gray with iron. The colors are stronger in autumn than in spring.

SHUNGIKU or edible chrysanthemum (*Chrysanthemum coronarium*) is an annual herb cultivated as a vegetable in Japan. All parts are scented. If you use boiled leaves, it dyes a beautiful green color with copper, dark brown with iron and brown with aluminum.

GENNOSHOKO or geranium (*Geranium thunbergii*) is a popular perennial wildflower. It is used as a traditional medicine for stomachache and diarrhea. Gather the leaves from summer to fall for dyeing. It gives yellow-brown with wood-ash juice and alum, liver color with copper and brownish-gray with iron.

PHOTOGRAPH BY AUTHOR

Delicate gradations of color can be produced in silk dyed with hops by changing mordants.

Kusagi-zome or *Kusagi* Dyeing

This is a special technique for obtaining beautiful blue dyes from the berries of *kusagi* or glorybower (*Clerodendron trichotomum*). Kusagi is a deciduous shrub growing 6 feet to 15 feet tall. Its opposite leaves have a bad smell, but the white flowers, borne from August through September, have a lovely fragrance. In the fall it bears blue berries with reddish violet calyxes. You can get a beautiful light blue color from the dyes without using a mordant, and a sober gray color from the calyxes. The blue dye works better on silk than on wool.

For dyeing, gather the jewel-like berries in October and separate them from the star-shaped calyxes. Use enough berries to equal two to three times the weight of the silk you want to dye. Put the berries in a nylon stocking and crush them thoroughly to extract the juice. Add the berry juice and the silk to a pan of water, heat gently to 85 degrees C (185 degrees F) and simmer for 45 minutes. To promote oxidation of the dye solution, add 0.5% by weight of hydrogen peroxide. Let the silk cool and rinse it thoroughly, then dry it in the shade. Redyeing the same silk with additional dyebaths gives a darker blue color. ❖

GROW YOUR OWN INDIGO

Rita Buchanan

Indigo is produced in the leaves of several plants, but most are tender tropicals. The three indigo sources that I've grown and used are woad (*Isatis tinctoria*), indigo (*Indigofera suffruticosa*) and dyer's knotweed or Japanese indigo (*Polygonum tinctorium*).

Woad

Woad is commonly planted in herb gardens, and has spread as a weed along roadsides and in agricultural fields in the United States. Some states list it as a "noxious weed", although it's much less invasive than bindweed, Russian thistle, or many other weeds. The responsible way to grow woad is to cut down the flower stems before the seeds mature; that done, it won't spread in your own or your neighbors' gardens.

To raise a supply of woad for dyeing, sow seeds directly in the garden at the time when daffodils are blooming. Space seeds a few inches apart and cover them one-half inch deep, and as the seedlings begin to grow, thin to one plant per square foot. Woad is a biennial, and grows a rosette of slightly hairy leaves the first season. Single plants can reach the size of a dinnerplate by midsummer. Harvest leaves for dyeing by plucking handfuls from around the bottom of each plant. More leaves will grow up in the center, and you can harvest from the same plants several times between July and October. Under good conditions, each woad plant may produce eight ounces of leaves over the season.

Usually I dig up all my woad plants before winter and toss them on the compost pile. Every few years, I leave two or three plants to winter over. Woad's no good for dyeing in the second year; all the potential indigo is coverted into another compound that triggers blooming. In early spring, each plant sends up one or more three-feet-tall stems of bright yellow flowers. After flowering, the rosette of leaves withers and dies, and hundreds of flat, papery, purple-brown fruits about one-half inch long dangle like earrings from the branched stems. I gather these before they start to blow away, to renew my seed supply.

Indigo

A shrubby perennial native to the New World tropics, indigo (*Indigofera suffruticosa*) can be grown as an annual crop where the growing season includes at least four months of daytime temperatures in the 80s or 90s, or you can grow a single plant in a six-inch flowerpot on a sunny windowsill just for fun. When the ground is comfortably warm in the spring, soak the hard little seeds in water overnight, then sow them one-half inch deep and three inches apart. Or start them indoors six to eight weeks earlier, and transplant them to the garden after all danger of cold weather is past. Indigo seedlings tolerate transplanting, but they sulk and stop growing if exposed to temperatures below about 45 degrees F. Thin or transplant seedlings to a spacing of eight inches apart. Indigo has erect stems three feet to four feet tall with short branches, and compound leaves two inches to three inches long with nine to 17 paired leaflets. You'll have to look closely to notice the delicate sprays of tiny pinkish-bronze pealike flowers. For dyeing, pick just the leaves, preferably during hot, dry weather and as the plant is beginning to bloom. Omit the stems; they are bulky and don't contribute any pigment to the dyepot. Harvested a few times at two-week intervals, a half-dozen indigo plants can provide one to two pounds of fresh leaves in a season.

Japanese Indigo

Dyer's knotweed or Japanese indigo (*Polygonum tinctorium*) is a tender annual native to eastern Asia and Japan and used there for centuries as an indigo source. You may be familiar with other species of knotweeds; many are common as garden flowers or weeds. Only dyer's knotweed is useful as an indigo source. It's *never* found wild in this country — you have to raise your own from seed. I sow seeds indoors six weeks before the last frost, and transplant the seedlings into the garden when the soil and weather are warm. I set them about 12 inches apart. Dyer's knotweed has smooth stems with swollen joints and simple alternate leaves about four inches long, and grows two feet to three feet tall. For dyeing, pick or strip the green leaves off the stems, starting from the base of the plant. You can pick half the leaves off a plant at one time without hindering its growth; in fact, the more leaves you pick, the more that grow. A half-dozen plants can yield more than two pounds of leaves between mid-summer and killing frost. Dyer's knotweed is quite showy in the fall, when it bears slender clusters of tiny pink flowers. Protect the plants with a blanket or sheet when the first light frosts come in fall, to keep them alive and give the seeds a chance to ripen. When the chaffy floral bracts have turned from pink to tan, gather the seed clusters and thresh out the small black seeds to save.

Dyeing with Fresh Indigo

Using indigo is different from dyeing with other plants, but it isn't any more difficult. I use the following method to dye with woad, *Indigofera suffruticosa*, or dyer's knotweed. From start to finish, the process takes less than eight hours. Like

Wood, *Isatis tinctoria*, growing in a row in a dyer's garden.

Here a dyer is shown pouring off an indigo solution.

baking bread, you can do other things in between the key steps.

Start by gathering fresh leaves. Pigment yield varies among these plants. I've gotten good color at plant:fiber ratios of 3:1 for *I. suffruticosa*, 4:1 for dyer's knotweed and 8:1 for woad. Weigh the leaves when you pick them, and weigh out an appropriate amount of dry yarn to dye at the same time. Stuff the leaves into a heat-resistant container, leaving little airspace. I put about a pound of leaves in a one-gallon container. There are two ways to take the next step, extracting the pigment. In the case of woad, pour almost-boiling water over the leaves to scald them. Use the minimum amount of water needed to cover the leaves. Cover the container and let it sit about an hour, then squeeze all the fluid out of the leaves and throw them away. Strain and save the warm, reddish brown liquid. With *I. suffruticosa* or dyer's knotweed leaves, add just enough hot tap water to cover the leaves, cover the container, set it down into a larger kettle of water (like using a double boiler) and heat it to 160 degrees F. over a period of two hours. You can heat it more slowly than that, but don't do it more quickly. Squeeze out the leaves and throw them away. Strain and save the warm, dark fluid. The colored fluids contain a precursor of indigo called indoxyl.

Regardless of which plant you started with, from here on the process is the same. *Don't* dilute the dyebath with any more water — you want it to be as concentrated as possible. Add a tablespoon of baking soda (sodium bicarbonate) per gallon of fluid, to make the solution alkaline. Then, pour the fluid back and forth from one container to another. As the indoxyl reacts with oxygen in the air, it

PHOTOGRAPH BY AUTHOR

Wool yarn dyed with *Polygonum tinctorium* is shown next to the plant itself.

Indigo-dyed cloth made of wool dyed, spun and woven by the author.

will change to a dark blue-green or blue-brown color. Keep pouring it back and forth for about several minutes.

To use the indigo for dyeing, it must first be converted from the blue form into a yellowish form called indigo white. Traditional methods of achieving this transformation involved adding stale urine and keeping the mixture warm for several days or weeks of smelly fermentation. Modern methods substitute chemical powders for the urine, and require only a few hours. I add either one tablespoon of Spectralite™ (available from dye suppliers) *or* two tablespoons of RIT Color Remover™ (available at grocery stores) per gallon of fluid. Sprinkle the powder on the surface of the dyebath and let it sink down by itself. For best results, keep the dyebath at about 100 degrees to 120 degrees F. at this time; I like to set the dyepot in a vessel filled with very hot water. Meanwhile, put your yarn to soak in warm water. (No mordant is required with indigo.) After an hour or more, when the dyebath has changed to a yellow-tan color, you can add the wet yarn, carefully lowering it down below the surface. Leave it to soak for 20 min. to an hour, then gently draw it out of the dyebath. The yellow color turns to blue as the indigo reacts with oxygen in the air. Let the yarn dry in the air for as long as it soaked in the bath. If you want to intensify the color, repeat the soaking and airing two or more times. After the final airing, wash and rinse the yarn.

Our ancestors knew plants well, as shown by the detail from the "Unicorn" tapestries (shown on page four), woven about 1500. All the colors in the tapestry were made with dyes from plants. Five hundred years later, they are still subtle and lovely. ❖

KUSAKI-ZOME JAPANESE GRASS AND TREE DYES

M. JOAN LINTAULT

Traditional Japanese resist-dyeing techniques were developed over a period of centuries. They produce beautiful colors and patterns, but the processes are time consuming, precise and labor intensive. Although threatened with replacement by modern techniques and synthetic dyes, the old dyes, called kusaki-zome, are still used today because of the Japanese appreciation for hand-crafted items. This technique was named, researched and developed by Akira Yamazaki in 1958.

The dyestuffs can be purchased from dye or herb suppliers, grown in the garden or foraged in the fields. Dye can be obtained from plant leaves, roots, bark, flowers, insects, fruits and berries or nut hulls. Most plants will yield some color when boiled and processed for dye, but only a few give deep, bright and long-lasting colors. For best results, leaves should be gathered in early summer, flowers just after blooming, whole plants when in bloom, fruit when well ripened, roots when the plant has died down in the fall, and bark in the spring. Plant parts can be dried out of direct sunlight and stored in a dark place for future use.

The Kusaki-zome method of dyeing involves making highly concentrated dye solutions from the plant materials. These solutions can be stored in the refrigerator until needed.

The dyes are extracted by a series of boilings. A general formula is to start with about four ounces of fresh or dried dyestuff in about three cups of hot water. Use an enamel, stainless steel or glass pot. Boil uncovered until less than half the liquid remains. Strain the liquid through a fine-mesh nylon fabric or stainless steel sieve. Return the same dyestuff to the pot, add more hot water and repeat. Save the results of each extraction separately. Continue the series of boilings until all the color is removed from the material. This may require as few as one extraction for petals and grasses or as many as 10 extractions for wood and bark. Combine the liquid from the first half of the extractions. These combined will give the darkest color. The liquid combined from the second half will give lighter colors. If desired, further boiling will reduce the combined liquid to make a very concentrated dye.

Extract and store all the dye that you will be using for one project at the same time. Each batch of natural dye is different, even if you use the same kind of plant, often because of variations in the growing conditions.

After boiling, let the dye stand overnight to settle any fine particles. Pour off and save the upper liquid and throw away the sediment. The dye can be stored in the refrigerator. Press a sheet of plastic against the surface of the liquid to seal it from air, and put a tight lid on the container. Since the dye is a natural substance, it is liable to mold if not used promptly. If mold forms on the surface, remove it carefully and test to see if the color remains satisfactory.

Using the Dyes

Kusaki-zome dyes can be applied by immersing the yarn or fabric in the dye liquid; or by painting, printing or stenciling the dye color directly onto a piece of fabric. Immersion dyeing is the process more familiar to many dyers, so I'll describe the general procedure for direct dyeing instead. These dyes can be used on any natural fiber: silk, wool, cotton, linen, ramie or viscose rayon.

First wash the fabric to remove all commercial sizing, then stretch it tightly on a frame or use tenter hooks (shinshi). Before dyeing, it is necessary to prepare the surface of the fabric by applying a sizing of soybean milk, which increases dye absorption, prevents wicking and streaking. You can use powdered soy milk from the health food store, mixed according to the directions on the label. Paint it sparingly on both sides of the fabric with a brush. Dry the soy milk-treated fabric promptly in the sun or in a warm room, so that it doesn't sour.

Adding thickener to the dyes helps keep them from wicking when applied to the fabric. A high viscosity low solids sodium alginate, a gum derived from seaweed, is the best thickener to use. It is available as a powder sold by dye suppliers. Mix a teaspoonful of sodium alginate with one-half cup water in a blender and let stand for at least an hour to thicken. The thickened alginate can be stored in the refrigerator and added to the dyes when necessary. Deciding how much thickener to add to the dye depends on the type of fabric, the dyestuff, and how many applications of dye will be used. From 10% to 50% are the usual amounts. Experiment on a scrap to see what works best for your purposes.

Most dyes require the presence of metallic salts, called mordants, to form a complete bond with the fibers. The Japanese use mordants of an acetic nature rather than a sulfuric acid nature because they believe that sulphate chemicals tend to remain on the fibers even after washing, causing color change or damage to the fibers. The mordants used are aluminum acetate, chromium acetate, cupric acetate, ferrous acetate, potassium dichromate and sodium stannate. There are other chemicals used as dye assistants such as tannic acid.

The formula for each mordant is different. Aluminum acetate is a good all-purpose mordant to begin experimenting with. It gives clear, delicate colors. Mix eleven level teaspoonfuls of powdered aluminum acetate in one quart of warm water.

Paint the thickened dye on the fabric freehand or use a stencil. Let it dry. Then apply the mordant by brushing it carefully onto the dry fabric over all the dyed areas. Making alternate applications of dye and mordant will deepen and strengthen the colors. Be sure that the fabric dries completely after each application. Repeat as many times as needed until the desired shade of color is obtained. To change colors, different dyes may be applied one over the other, or mixed together before using.

Sappan wood and madder require slightly different formulas. After treating the fabric with mordant, these dyes on cotton require the application of tannin to the fabric in layers alternating with dyestuff and mordant.

After you have finished painting, the fabric must be steamed to complete the bond between dye, mordant and fiber. Steam pieces of fabric by layering them between sheets of newspaper and placing the stacked bundle on a rack in a large canning pot or steamer. Make sure no two pieces of fabric touch each other or the sides of the pot, or staining will occur. Steam the fabric for about 45 minutes.

Wash the fabric in a mild soap such as Woolite, handling it gently at all times. Rinse several times in tepid water. Do not rub or wring the fabric. Roll it in a towel to remove the excess water or spin out the water in the washing machine. Dry in the open air, protected from direct sun. ❖

These bolts of kimono fabric were dyed naturally.

PHOTO BY M. JOAN LINTAULT

MORDANT SAFETY

Rita Buchanan

Traditionally, dyers took mordant compounds for granted, but contemporary dyers often express concern about safe handling and disposal of these chemicals. It's hard to find reliable and up-to-date answers to questions about mordant safety. To compile this article, I had to draw upon a variety of sources including chemical lab manuals, industrial safety handbooks, medical accounts of accidental poisoning and directories of toxic substances. Given below are quick summaries of what I found. Learning more about the metallic salts we dyers use as mordants reminded me to treat them with respect but reassured me that careful dyers are not likely to harm themselves or their environment.

For your own safety and the safety of those around you, always follow these general rules. Keep mordants in labeled containers. Store them out of reach when not in use. Be careful not to inhale dust-fine particles when measuring mordants. Wear rubber gloves and avoid skin contact with mordant powders or solutions. Wash mordanted yarns thoroughly with soap and water. Never use the same containers and implements for mordanting, dyeing and cooking. Use separate equipment for dyeing, label it clearly and store it away from the kitchen.

Proper disposal of dye wastes depends on the mordants and quantities involved. Remember that the purpose of mordanting is to fix the metal to the fiber; there shouldn't be much mordant left over. Measure accurately and use the minimum required. There's little danger from disposing occasional small batches of used alum, tin, iron or copper mordant baths onto the ground (always keep a distance from wells or waterways); chrome may pose a hazard. If you do a lot of dyeing and must frequently dispose of large quantities of waste, ask your local water or waste-control agency for advice. Many communities now sponsor an annual hazardous waste collection day, when households can bring cleaning solvents, automotive fluids, pesticides, dry-cell batteries, photographic chemicals and hobby supplies such as mordants for special handling and disposal.

Background on the Most Common Mordants

Aluminum Aluminum is a good mordant for wool and silk and other protein fibers, and is used in combination with tannins for cotton and flax and other cellulose fibers. The aluminum salt most frequently used as a mordant is potassium aluminum sulphate, or alum, a white powder. (Another kind of alum, ammonium aluminum sulphate, is less desirable.) Using aluminum dye vessels has some effect in brightening dye colors, but not enough aluminum dissolves to serve as a mordant.

Alum is generally recognized as safe to handle. It is an ingredient in deodorant, foot powder and styptic pencils. Ingesting a small amount causes dryness and a puckering sensation in the mouth and throat. Larger doses cause vomiting or diarrhea, but few if any fatalities have been attributed to alum.

Gardeners use alum as a soil acidifier for acid-loving plants such as rhododendrons, azaleas, camellias or blueberries. Used alum mordant baths can be poured on the ground, but don't overdo. Repeated application can build up concentrations of aluminum which are harmful to plants.

TIN Stannous chloride (tin chloride) is the tin compound used in mordanting. It occurs as white crystals or flakes. Tin brightens colors, particularly reds, oranges and yellows. Too much tin makes wool or silk brittle and fragile. To avoid this, some dyers use just a pinch of tin, added at the end of the dyeing, on wool premordanted with alum. Dyeing in tin cans (which are really tin-plated steel cans) or tin-plated pots doesn't dissolve enough tin to suffice as a mordant.

Tin is generally considered non-toxic to humans. The human body tends not to absorb tin; for example, the small amount dissolved in canned foods is readily excreted. Ingesting a sip of a tin mordant solution would cause nausea and diarrhea, but is unlikely to be fatal. On the other hand, be aware that chlorine is a poisonous gas. When you dissolve tin chloride in water, it gives off fumes which can irritate your eyes, nose, and skin. Work in a well ventilated area or use an exhaust fan to carry away the chlorine fumes.

It is safe to dispose of occasional batches of tin mordant by pouring the liquid on the ground. The chlorine will evaporate, and the remaining tin is no more harmful than tin cans.

IRON Traditional dyers used iron-rich muds or water as a mordant, or simmered their dyebaths in iron pots. These methods are still effective, but modern dyers often use iron sulphate, sometimes called copperas, for an iron mordant. Iron sulphate is a greenish powder. As a mordant, iron is used to make dark colors, often producing dark greens, grays or even black. Unfortunately, using iron often weakens fibers, and iron-mordanted yarns become brittle and deteriorate over time.

Most people feel comfortable with iron because it's a familiar metal, and many forms of iron, including iron oxide, or rust, are quite harmless. The effect of iron sulphate, however, is critically dependent on the dose. Iron sulphate is the same form of iron that's used in iron tablets. Iron is an essential constituent of the human body, necessary for hemoglobin formation. An adult human's body contains about 3.5 grams of iron, of which one to two grams are lost daily. But in children and adults, as little as 5 to 10 grams of iron sulphate can be fatal. Acute poisoning from accidental ingestion of iron tablets affects 2000 children a year: Only aspirin poisonings occur more frequently. Iron poisoning is dangerous because iron is readily absorbed, not excreted, and the concentrations build up to high levels. The symptoms are severe gastric disturbance, followed by headache, delirium, convulsion and death. The bottom line is this: Treat iron sulphate with as much respect as you give other mordants.

Plants need iron, and gardeners use iron sulphate to correct iron deficiencies and also to acidify the soil. Broadleaf evergreens are especially likely to show iron deficiency — their leaves get yellow, and growth is stunted. It's safe to dispose of occasional batches of iron mordant by pouring the liquid around the base of hollies, rhododendrons and other evergreens.

COPPER As a mordant, copper is used like iron, to darken or "sadden" the dye colors. Copper is less harsh on the fibers than iron. Simmering dyebaths in a copper pot adds enough copper to make a difference in the color, and can be used in combination with alum mordant. Dyers without a copper pot use copper sulphate, also called blue vitriol, as a copper mordant. Copper sulphate forms brilliant blue crystals.

The human body needs a trace of copper; it functions in enzymes and in the nervous system. However, ingesting even a swallow of copper sulphate solution causes violent vomiting, generally within 15 minutes. Too much copper causes severe, irreversible damage to the liver and kidneys. Ingesting six grams of copper sulphate is likely to be fatal to an adult. Also, contact with copper sulphate may cause allergies and irritation of the skin, eyes and upper respiratory tract.

Handled with respect, copper sulphate is a very useful chemical. Gardeners sometimes add it to peat or muck soils which are deficient in copper, necessary for plant growth. More commonly, copper sulphate is used as a fungicide to control blights, mildews and leaf spots. The United States as a whole uses over 35,000 tons of copper sulphate annually — for agricultural fungicides and fertilizers, in metal finishing and wood preservatives, to control root growth in sewers, and to kill algae in water supplies. It's safe to dispose of occasional batches of copper mordant by pouring the liquid onto the ground.

CHROMIUM The form of chromium most frequently used as a mordant is potassium dichromate, an orange powder. It gives bright colors, and makes wool feel soft and silky. Chromium occurs in several forms that pose different health risks. The hazard of exposure to different chrome compounds is a subject of much controversy.

As adults, our bodies contain approximately six grams of chromium. It is essential for the production of insulin, and also affects protein synthesis. The food in a normal diet supplies adequate chromium, mostly in a form which passes through the body and is excreted. However, ingesting as little as 0.5 grams of potassium dichromate — the form of chrome used for mordanting — can cause kidney failure, coma and convulsions. It's estimated that six to eight grams of potassium dichromate would be a fatal dose for an adult, although there are no records that this kind of poisoning has ever occurred.

It's not well known that chromium is one of the most common causes of skin allergy, especially among industrial workers. Exposure to the chromium in cement dust, primer paints, auto anti-rust coating, detergents and bleaching agents can cause skin rashes and skin ulcers — recessed spots that heal slowly, although they are not painful and are not permanent. Inhaling the dusts and fumes of chromium compounds can cause respiratory problems and may cause lung cancer. Most of the hazards of chrome exposure have been reported from sites where factories have deposited huge piles — up to two million tons at one site in New Jersey — of chrome wastes. I could find no reports of serious symptoms experienced by hobby dyers who use an ounce or two of potassium dichromate now and then.

For guidelines on disposing of a used chrome mordant bath, contact your local authorities. Some municipalities would consider it a hazardous waste requiring special disposal, and others would not.

FURTHER READING

Arena, Jay M., ed. *Poisoning*, 5th ed. Springfield, IL: Charles C. Thomas, 1986.

NIOSH: *Pocket Guide to Chemical Hazards.* National Institute for Occupation Safety and Health, Cinncincinnati, Ohio, 1985.

Proctor, N.H. and J.P. Hughes. *Chemical Hazards of the Workplace.* Philadelphia: J.B. Lippincott Co., 1978.

GETTING STARTED WITH DYEING

RITA BUCHANAN

Little equipment is needed for dyeing, but mark each piece with indelible labels, and *don't* use the same equipment for cooking. Store the equipment in a safe place, out of reach of children. You'll need:

- a stainless steel or enamel pot that can hold several gallons of liquid;
- wood or metal rods for stirring the dyebath and handling the fibers;
- scales for weighing fibers, dyestuffs and mordants;
- a set of measuring spoons and cups;
- a wire mesh strainer;
- a candy, dairy or scientific thermometer to use in the dyebath;
- rubber gloves;
- a stove, hotplate or other heat source;
- a well-ventilated work area.

Wool is the easiest natural fiber to dye. It absorbs and retains colors well, and isn't too much trouble to handle. If you're new to dyeing, I'd recommend that you start with white or "natural" wool yarn. First wind it into skeins about a yard in circumference, and weigh them with a postal scale — two to four ounces makes a good size skein. Tie the two ends of the yarn together in a bow or knot, leaving tails a few inches long, and use three or four pieces of string to tie loops around the skein to prevent its tangling. For the dye to penetrate evenly, the yarn must be clean, so remove any oils or finishes by washing it in warm water and liquid dishwashing detergent. Then rinse it in warm clear water.

Many dyes give better results if the yarn is treated with a small amount of a metallic salt, called a mordant. Mordants based on aluminum, tin, chrome, copper and iron are available through crafts stores and chemical supply houses. Handle mordants with care, and keep the labelled containers in a safe place out of the reach of children. Using different mordants increases the range of colors that you can get from a single dye plant,

PHOTOGRAPH BY AUTHOR

Yarns dyed from plants grown in the author's garden.

but I use alum (potassium aluminum sulphate) far more often than I do any of the others. It is nontoxic and readily available, and it doesn't make the wool harsh as some other mordants do.

To mordant wool, use a big pot and allow at least four gallons of water per pound of wool. For one pound of wool, measure out four tablespoons of alum plus four teaspoons of cream of tartar; *or* one tablespoon of chrome; *or* two tablespoons of copper; *or* two tablespoons of iron; *or* two teaspoons of tin. Dissolve the measured amount of mordant in a cup of boiling water, add it to the pot of water and stir well. Add wet wool, heat slowly to a simmer and simmer for an hour. Set the pot off the heat and let it cool for several hours until it is lukewarm. Remove the wool and rinse well, then dye it immediately or dry it to store for future use.

The depth of color depends on the kind of plant and the relative quantities of plant and yarn. Some plants give much more color than others. I usually start with a 4:1 ratio of fresh plants to wool (by weight), because I like rich colors. Using less dyestuff gives lighter colors. To make a dyebath, put the plants in the dyepot and cover it with water. Heat to just below boiling and simmer for at least an hour. With fresh soft plant parts such as flowers and leaves, I usually strain off the dye liquid right after boiling and proceed with the dyeing. Hard parts such as roots, bark or nut hulls release more color if you let the pot cool overnight and drain off the liquid the next day.

Finally, place the mordanted wool in the dyebath. Bring it slowly to a simmer and maintain for an hour. Remove at once or let the yarn cool in the dyebath overnight for darker colors. Wash and rinse the yarn until the color stops bleeding, then squeeze out excess moisture and spread or hang the yarn to dry. ❖

A NATURAL DYEING BIBLIOGRAPHY

ANNE BLISS

Adrosko, Rita J. (1971) ***Natural Dyes and Home Dyeing***. New York: Dover Publications, Inc. (154 pages) and Furry, Margaret S. and Bess M. Viemont ***Home Dyeing with Natural Dyes.***
This combined book features historical dye methods and plants used by American colonists.

Bemiss, Elijah. (1973) ***The Dyer's Companion***. New York: Dover Publications, Inc. (313 pages)
This book is a complete reprinting of Bemiss's original 1815 text on natural dye methods and how to produce over 100 shades on wool, linen, and cotton.

Bliss, Anne. (1981) ***A Handbook of Dyes from Natural Materials***. New York: Charles Scribner's Sons. (180 pages)
Recipes from across the U.S. are presented in cookbook format in this book with both black and white and color illustrations. Some history as well.

Bliss, Anne. (1986) ***North American Dye Plants***. Boulder, CO: Juniper House. (288 pages)
The pocket or backpack sized format of this book make it ideal for foraging trips as well as easy to use in the dye area. 126 plants are illustrated with instructions for mordanting, dyeing.

Bronson, J. and R. (1977) ***Early American Weaving and Dyeing***. New York: Dover Publications, Inc. (204 pages)
Although this book was published nearly 200 years ago, most of the information is still useful to the home artisan. The dye recipes are interesting reading, and they work!

Buchanan, Rita. (1987) ***A Weaver's Garden***. Loveland, CO: Interweave Press, Inc. (230 pages)
Although this book includes more than just dye plants, the chapter on dyes from plants is a thorough introduction to growing traditional favorites such as

weld and indigo as well as to methods for using them.

Davidson, Mary Frances. (1981) ***The Dye Pot.*** Rt. 1, Gatlinburg, Tennessee: Author. (27 pages)
This booklet is a classic "how to" for dyers. Chock-full of useful, basic information, the book lists plants alphabetically with color possibilities obtainable by using various methods.

De Boer, Janet, Ed. (1987) ***Dyeing for Fibres and Fabrics***, rev. ed. Kenthurst, Australia: Kangaroo Press Pty. Lts. (112 pages)
Although this book is intended primarily for Australian dyers, the information provided by the 21 dye experts who have written on techniques and uses for dyes is useful around the world.

Fraser, Jean. (1989) ***Traditional Scottish Dyes.*** Edinburgh, Scotland: Canongate Publishing Ltd. (106 pages)
This book gives recipes for 75 plants, most available in North America. Basic information about methods, parts of plants to use, and color yields from various mordants is provided.

Hurry, Jamieson B. (1973) ***The Woad Plant and Its Dye.*** Clifton, N.J.: Augustus M. Kelley Pub. (328 pages)
Reprinted from Oxford University Press's original 1930 edition, this book presents a comprehensive history and botanical information about the woad plant. Recipes and methods are given.

Kierstead, Sallie Pease. (1972) ***Natural Dyes.*** Boston: Branden Press Publishers. (104 pages)
Basically a "how to" book, this text gives general information about plants and various methods to use with them.

Miller, Dorothy. (1984) ***Indigo from Seed to Dye.*** Aptos, CA: Indigo Press.
This book is a delightful essay on the virtues of indigo, and it provides good background on Japanese use of indigo-bearing plants to produce traditional as well as contemporary textiles. The author tells how to set up a vat and use it, and even supplies a coupon in the book that may be exchanged for a free packet of "indigo seeds" from *Polygonum tinctorium.*

Rice, Miriam and Dorothy Beebe. (1980) ***Mushrooms for Color.*** Eureka, CA: Mad River Press. (154 pages)
Here's an interesting book that not only tells how to collect mushrooms, but how to use them to produce wonderful colors on silk and wool.

Sandberg, Gosta. (1989) ***Indigo Textiles Technique and History.*** Asheville, NC: Lark Books. (184 pages)
This English-language version of the original Swedish book published in 1986 is a gem. Delightful history, beautiful colored photos, and the best assortment of uses for indigo I've seen in a book for home dyers are gathered together in this volume.

Young, Stella and Nonaba G. Bryan. (1978) ***Navajo Native Dyes.*** Palmer Lake, CO: Filter Press. (unpaged)
Originally published by the U.S. Bureau of Indian Affairs, this little book describes methods developed to teach the Navajo how to use native dyestuffs found on the reservation. Although many of the original methods had been lost in the 100 years previous to the 1940 issue, some dyers were still practicing. ✣

Sources of Dye Supplies and Dye Plants

Mail-Order Suppliers of Dyestuffs and Mordants

Alliance Import Co, 1021 R. Street, Sacramento, CA 95814; (916) 920-8658. Attn: Roger Wolfe.

Batik and Weaving Supplier, 393 Massachusetts Ave., Arlington, MA 02174; (617) 646-4453.

Brooks and Flynn, P.O. Box 2639, Rohnert Park, CA 94928; (800) 822-2372; in CA (800) 345-2026.

Bryant Laboratory Inc., 1101 Fifth St., Berkeley, CA 94710. (415) 526-3141.

Coupeville Spinning and Weaving Shop, P.O. Box 520, Coupeville, WA 98239; (206) 678-4447.

Creek Water Wool Works, P.O. Box 176, Salem, OR 97308; (503) 585-3302.

DBP Unlimited, P.O. Box 1344, Duarte, CA 91010; (818) 357-8677.

Diana's Designs, 7011 Spieth Road, Medina, OH 44256; (216) 722-2021.

Earth Guild, One Tingle Alley, Asheville, NC 28801; (800) 327-8448.

The Mannings, P.O. Box 687, East Berlin, PA 17316; (717) 624-2223.

Rio Grande Weaver's Supply, PO. Box 2009, Taos, NM 87571; (505) 758-0433.

Testfabrics, Inc., P.O. Box 420, Middlesex, NJ 08846; (201) 469-6446.

Mail-Order Suppliers of Dye Plants and Seeds

Flowery Branch, PO Box 1330, Flowery Branch, GA 30542

Forestfarm, 990 Tetherow Rd., Williams, OR 97544

Goodwin Creek Gardens, Box 83-C, Williams, OR 97544

Dorothy Miller, 5950 Fern Flat Road, Aptos, CA 95003 (indigo seeds)

Native Seeds/SEARCH, 3950 West New York Drive, Tucson, AZ 85745 (indigo seeds)

Richter's Herbs, Goodwood, Ontario L0C 1A0 CANADA

Sandy Mush Herb Farm, Rt. 2, Surrett Cove Rd, Leicester, NC 28748

Well Sweep Herb Farm, 317 Mt. Bethel Road, Port Murray, NJ 07865

World Seed Service, PO Box 1058, Redwood City, CA 94064